AF784683

CONTENTS

Opening statements: Page
Senator Collins 1
Senator Lieberman 2
Senator Warner 19
Senator Levin 21
Senator Dayton 25
Senator Akaka 29
Senator Pryor 33
Senator Lautenberg 38
Senator Carper 41

WITNESS

THURSDAY, OCTOBER 20, 2005

Marty J. Bahamonde, Director, External Affairs, Region One, Federal Emergency Management Agency, U.S. Department of Homeland Security:
Testimony 4
Prepared statement 51
Post-hearing questions and responses 56

APPENDIX

Exhibit No. 1 60
Exhibit No. 2 61
Exhibit No. 3 62
Exhibit No. 4 63
Exhibit No. 5 64
Exhibit No. 6 65
Exhibit No. 14 66
Exhibit No. 22 67
Exhibit (Senator Lieberman) 68

S. Hrg. 109–482

HURRICANE KATRINA IN NEW ORLEANS: A FLOODED CITY, A CHAOTIC RESPONSE

HEARING

BEFORE THE

COMMITTEE ON HOMELAND SECURITY AND GOVERNMENTAL AFFAIRS UNITED STATES SENATE

ONE HUNDRED NINTH CONGRESS

FIRST SESSION

OCTOBER 20, 2005

Printed for the use of the
Committee on Homeland Security and Governmental Affairs

U.S. GOVERNMENT PRINTING OFFICE
24–442 PDF WASHINGTON : 2006

For sale by the Superintendent of Documents, U.S. Government Printing Office
Internet: bookstore.gpo.gov Phone: toll free (866) 512–1800; DC area (202) 512–1800
Fax: (202) 512–2250 Mail: Stop SSOP, Washington, DC 20402–0001

COMMITTEE ON HOMELAND SECURITY AND GOVERNMENTAL AFFAIRS

SUSAN M. COLLINS, Maine, *Chairman*

TED STEVENS, Alaska
GEORGE V. VOINOVICH, Ohio
NORM COLEMAN, Minnesota
TOM COBURN, Oklahoma
LINCOLN D. CHAFEE, Rhode Island
ROBERT F. BENNETT, Utah
PETE V. DOMENICI, New Mexico
JOHN W. WARNER, Virginia

JOSEPH I. LIEBERMAN, Connecticut
CARL LEVIN, Michigan
DANIEL K. AKAKA, Hawaii
THOMAS R. CARPER, Delaware
MARK DAYTON, Minnesota
FRANK LAUTENBERG, New Jersey
MARK PRYOR, Arkansas

MICHAEL D. BOPP, *Staff Director and Chief Counsel*
DAVID T. FLANAGAN, *General Counsel*
JOYCE A. RECHTSCHAFFEN, *Minority Staff Director and Counsel*
LAURIE R. RUBENSTEIN, *Minority Chief Counsel*
TRINA DRIESSNACK TYRER, *Chief Clerk*

HURRICANE KATRINA IN NEW ORLEANS: A FLOODED CITY, A CHAOTIC RESPONSE

THURSDAY, OCTOBER 20, 2005

U.S. SENATE,
COMMITTEE ON HOMELAND SECURITY
AND GOVERNMENTAL AFFAIRS,
Washington, DC.

The Committee met, pursuant to notice, at 9:33 a.m., in room SD–342, Dirksen Senate Office Building, Hon. Susan M. Collins, Chairman of the Committee, presiding.

Present: Senators Collins, Warner, Lieberman, Levin, Akaka, Carper, Dayton, Lautenberg, and Pryor.

OPENING STATEMENT OF CHAIRMAN COLLINS

Chairman COLLINS. The Committee will come to order. Good morning.

Today we will hear from the first FEMA employee who was on the ground in New Orleans immediately before, during, and after Hurricane Katrina. Marty Bahamonde is a career employee at FEMA who has responded to numerous natural disasters, including hurricanes in Florida and the massive earthquake in Iran in 2003. Senator Lieberman and I met Mr. Bahamonde during our tour of the Gulf Coast States last month. At that time he gave us a sobering assessment of the response to Katrina at all levels of government.

Contrary to former FEMA Director Michael Brown's testimony before the House Select Committee, Mr. Bahamonde told us that he was the only FEMA official in the area before and during the hurricane.

On Friday, August 26, FEMA headquarters asked Mr. Bahamonde to position himself near ground zero of the looming storm. He arrived in New Orleans late Saturday night, and on Sunday morning he made his way to the city's Emergency Operations Center (EOC), which happened to be across the street from the Superdome. He then began providing updates to FEMA officials in Washington about the situation in New Orleans. He made phone calls, wrote E-mails, and sent photographs.

Sunday noon, recognizing that the crowds at the Superdome were larger than the people at the EOC had anticipated, he notified FEMA officials in Washington of the developing problem. On Monday, he rode out the storm at the EOC and at a parking garage where he had a good view of the downtown and the Superdome. He provided on-the-scene updates about the catastrophic devastation the city was suffering. He was at the EOC at 11 a.m. on Monday when the people of the center were first notified that one of the levees was damaged.

Over and over, he alerted top FEMA officials that the situation was dire and that it was growing worse by the hour. He also briefed city officials on what he learned. Yet, inexplicably, his urgent reports did not appear to prompt an urgent response.

Following the storm, the already grim situation at the Superdome became a nightmare. The Superdome uncontrollably became a mass shelter for tens of thousands where providing such basics as sanitation became impossible and food and water were a constant struggle.

At 4 a.m. Tuesday, when the flood waters reached the Superdome, the logistical problems there grew exponentially. As Mr. Bahamonde reported in calls and briefings on Monday, Tuesday, and Wednesday, the crisis was exploding. In an E-mail to Director Brown on Wednesday, August 31, he described the situation as "past critical" and said that the Superdome was out of food and running out of water.

Within FEMA itself, there was a failure to coordinate effectively. Mr. Bahamonde tried desperately to fill the void created by poor planning and a sluggish response. His accounts from the scene, some of which are preserved in E-mail messages, describe heroic efforts of FEMA employees deployed to the Superdome after the hurricane. But his E-mails also describe a complete disconnect between senior officials and the reality of the situation on the ground. Mr. Bahamonde wrote a poignant message on Saturday, September 3, in which he saluted the work of many FEMA employees, but also stated, "The leadership from top down in our agency is unprepared and out of touch."

Mr. Bahamonde should be commended for his service at FEMA, and the information that he is providing this Committee will help us probe and better understand many troubling questions, such as how exactly did local, State, and Federal Governments react during the 16 hours between the report of the break in the levee and the time that the flood waters reached the Superdome.

As the Committee's investigation proceeds, we will examine further why critical information provided by Mr. Bahamonde was either discounted, misunderstood, or simply not acted upon. For now, I look forward with great interest to hearing his testimony.

Senator Lieberman.

OPENING STATEMENT OF SENATOR LIEBERMAN

Senator LIEBERMAN. Thanks, Madam Chairman.

Our witness today, Marty Bahamonde, was the first and, for too long, just about the only FEMA employee on the ground during Hurricane Katrina in New Orleans. He arrived before the storm hit, rode it out in the city's Emergency Operations Center, and ended up being one of the very small group of FEMA workers in New Orleans for a full 3 days after the storm. During that time, Mr. Bahamonde offered invaluable firsthand information to his FEMA superiors and was of great help to the thousands of people trapped in inhumane conditions at the Superdome.

I consider Marty Bahamonde's actions in New Orleans to be heroic, but the jolting fact is that he was an accidental hero. He is not a member of the FEMA Emergency Response Teams or FEMA Disaster Medical Assistance Teams or Search and Rescue Teams. Marty Bahamonde is a FEMA public affairs officer who was sent to New Orleans to advance the expected trip of then-FEMA Director Michael Brown to the city and to provide information before Mr. Brown's arrival to him about what was actually happening in New Orleans.

Mr. Bahamonde's story is powerful and it is moving, and I will leave it to him to tell it. But his story is also deeply troubling—in fact, ultimately infuriating, and raises very serious questions which this Committee's investigation must answer.

For instance, why after the National Weather Service so clearly predicted on Friday, August 26, and, in fact, before that Katrina would be a hurricane of Category 4 or 5, the big one long feared and likely to literally drown New Orleans, why after all those warnings would FEMA manage to get Marty Bahamonde, a public affairs officer, into New Orleans the next day, Saturday, August 27, but took until Tuesday, August 30, in the morning to send a single Disaster Medical Assistance Team and until Tuesday afternoon for a small four-person FEMA Emergency Response Team to arrive?

So you have got all these warnings of the storm coming. FEMA dispatches Mr. Bahamonde there on a previously mentioned assignment on Saturday. But the real help does not arrive until Tuesday after the Superdome is in humanitarian crisis and most of New Orleans is under water. That is an unbelievable and unacceptable turn of events.

In light of the fact that Mr. Bahamonde began sending reports to his superiors on Sunday, August 28, the day before landfall, of swelling crowds and significant humanitarian needs at the Superdome, why did FEMA send so few medical assistance and emergency response personnel to the scene?

In Mr. Bahamonde's testimony this morning, we are going to hear that on Monday, August 29, the day Katrina did make landfall, he reported to FEMA, first in late morning E-mails, then later in the day in phone conversations including one directly with Michael Brown, that some of the levees around New Orleans had broken and that, in his own words, 80 percent of New Orleans was under water. Why then have Secretary Chertoff as recently as yesterday, before a House committee, and Secretary Rumsfeld and General Myers said that the levees were not breached until Tuesday, when Marty Bahamonde was telling everybody on Monday that they had already been breached, or at least Secretary Rumsfeld, Secretary Chertoff, and General Myers indicating that they did not learn of the levees failing until Tuesday?

We are going to hear this morning how Mr. Bahamonde told everyone he could up to Michael Brown that the levees had broken early Monday morning and that New Orleans was literally drowning. And yet two of our high Cabinet officials say they did not know that until Tuesday.

This disconnect between what Marty Bahamonde will tell us he saw and told his superiors on Monday morning and what Secretaries Chertoff and Rumsfeld have said they did not know until Tuesday is beyond disturbing. It is shocking. Our investigation must find out what happened to Mr. Bahamonde's reports from

New Orleans. Who received them in the administration? What did they do with them? Do the answers to those questions help explain the excruciatingly slow response of the Department of Homeland Security, of FEMA, even of the Department of Defense, to Hurricane Katrina and the human suffering it left in its wake? Those are critical questions our investigation must answer.

The testimony we will soon hear from Mr. Bahamonde will leave us better informed, much better informed, but I believe also much more troubled by how the Federal Government responded to this hurricane.

Mr. Bahamonde, I thank you for your dedicated work before, during, and after Hurricane Katrina hit land and for your important testimony this morning. And I promise you that the Members of this Committee, will do everything we can to make sure that nothing like the chaos and suffering you witnessed with your own eyes and tried as hard as you humanly could to stop in New Orleans will ever happen in America again.

Thank you.

Chairman COLLINS. Thank you, Senator.

Mr. Bahamonde, I know that this has been an exhausting and difficult past 2 months for you, and I want to thank you sincerely on behalf of all the Members of this Committee, and both past and future victims of natural disasters for your willingness to come forward. Our goal is to identify what went wrong so that we can improve the system and ensure that mistakes are not repeated.

So I just want to publicly thank you so much for coming forward, and I would like to ask you to proceed with your testimony.

TESTIMONY OF MARTY J. BAHAMONDE, DIRECTOR, EXTERNAL AFFAIRS, REGION ONE, FEDERAL EMERGENCY MANAGEMENT AGENCY, U.S. DEPARTMENT OF HOMELAND SECURITY

Mr. BAHAMONDE. Thank you, Senator. Good morning, everyone. I appreciate the opportunity to appear before you today.

As you have already mentioned, I work for the Federal Emergency Management Agency (FEMA) as a public affairs officer for FEMA's Boston office, and previous to that worked in FEMA's Headquarters here in Washington, DC. I worked in New Orleans prior to and immediately following Hurricane Katrina and have spent the past 6 weeks working at the Joint Field Office in Baton Rouge. As you mentioned, I was the only FEMA employee deployed to New Orleans prior to the storm. I am here today to tell you what I experienced during the 5 days before and after Hurricane Katrina in New Orleans and specifically at the Superdome.

I have worked for FEMA for 12 years and have been a full-time employee since 2002. I have spent most of that time in the field, not behind a desk. I have responded to numerous hurricanes, earthquakes, floods, tornados, and fires. At the personal request of Under Secretary Brown, I deployed to Bam, Iran, in 2003 in support of the medical team that worked miracles in that city that was totally destroyed and resulted in more than 30,000 deaths. I have seen and lived through the worst that Mother Nature can provide, and I saw it this year in New Orleans and the Superdome.

Let me briefly explain why I was there. Since 2003, I have often been tasked by Under Secretary Brown and his staff to do advance work in preparation for or responding to large natural disasters. My assignments have included Hurricane Isabel in 2003, Hurricane Charley in 2004, and Hurricane Dennis this past June, and others. My responsibilities varied, but also included providing accurate, timely, and important information to FEMA's front office and Under Secretary Brown himself. On August 26, I was tasked by FEMA's front office to work advance wherever Hurricane Katrina was going to hit, which on Saturday appeared to be New Orleans.

I arrived in New Orleans Saturday night, August 27, and through the generosity of the New Orleans Emergency Operations Center (EOC), I was able to work in their office, and they provided me shelter during the hurricane. As you have already mentioned, the city EOC is located in City Hall, almost directly across from the Superdome.

On Sunday, August 28, the day before the hurricane hit, I met with the city staff at the EOC; I got to know the people in their roles at the EOC; and I developed my own plans for my own advance work, which included coordinating with the Coast Guard to arrange a flyover after the hurricane had passed. On Sunday, Katrina intensified to a Category 5 storm, with estimated winds of 175 miles per hour. I sensed a great deal of worry among everyone at the EOC, and, frankly, I was worried, too. My contact at FEMA Headquarters told me to leave New Orleans because it would be too dangerous. But like the thousands of other people left in New Orleans, the traffic jams leaving the city that morning prevented me from leaving as well.

The Superdome had been designed as a special needs shelter, but on Sunday, as thousands of residents were unable to evacuate, the Superdome became a shelter of last resort for anyone left in the city. By noon, thousands began arriving, and by midday, lines wrapped around the building. It was also at that time that I realized that the size of the crowd was becoming a big concern at the EOC. Terry Ebbert, the city's Homeland Security Director, made an announcement in the EOC that struck me. He asked the maintenance staff to gather up all of the toilet paper in City Hall and any other commodities they could find and immediately take them over to the Superdome. I specifically noted this because it told me that supplies at the Dome were going to be and were already a serious issue.

I was between the Superdome and the EOC throughout the day on Sunday. I took pictures of the crowds and sent them back to FEMA Headquarters, which were posted online. On Sunday evening, I was at the Superdome to do media interviews, and afterwards I met with the National Guard inside the Superdome to discuss a wide range of things, including the expected arrival of a FEMA Disaster Medical Assistance Team from Houston. The National Guard also told me at that time that they expected 360,000 MREs and 15 trucks of water to arrive later that night. But as the storm intensified outside, a series of critical events began to unfold. Instead of 360,000 MREs, only 40,000 arrived. Instead of 15 trucks of water, only 5 arrived. And the medical team did not arrive either.

Later that night, after most of the 12,000 evacuees had entered the Superdome, I returned to the EOC around midnight to ride out the storm. By early Monday morning, with the storm upon us, reports from throughout the city were moderately optimistic: Some low-level flooding, no levee breaks, and limited wind damage. But by 8 a.m., the nearest point of eye passage, the situation worsened. I could clearly see and reported back to FEMA Headquarters that morning that the Hyatt Hotel and many other tall buildings in downtown had suffered incredible damage, and I could see the roof peeling off the Superdome. I received numerous calls from FEMA Headquarters seeking confirmation of the situation on the ground.

At approximately 11 a.m., the worst possible news came into the EOC. I stood there and listened to the first report of the levee break at the 17th Street Canal. I do not know who made the report, but they were very specific about the location of the break and the size. And then they added it was "very bad." I immediately called FEMA's front office to relay the news. Their reaction was shock and disbelief.

As I passed on minute-by-minute information, I was always under the assumption that it would then be passed to Under Secretary Brown and others. I do not know if this ever happened. I also contacted FEMA Public Affairs, and I do know that an E-mail was sent to staff who was with Under Secretary Brown in Baton Rouge mentioning the levee failure and other critical information, including the mayor's concern at that time that an estimated 30,000 tourists were still in town occupying hotel rooms. I continued to provide regular updates to FEMA Headquarters throughout the day as the situation unfolded.

Later that afternoon, at approximately 5 p.m., I rushed over to the Superdome because I had been notified that a Coast Guard helicopter was able to take me for a short flyover so that I could assess the situation in the city and plan for Under Secretary Brown's visit the next day. My initial flyover lasted only 10 minutes, but even in that short amount of time, I was able to see that approximately 80 percent of the city was under water, and I was able to visually confirm the 17th Street Canal levee break. I was struck at that moment by how accurate the 11 o'clock call about the levee was.

About 15 minutes later, I went back up on a second Coast Guard helicopter for approximately 45 minutes, and during this flight I was able to get a true understanding of the impact of Katrina on New Orleans and the surrounding area.

Upon landing, at approximately 7 o'clock, I immediately made three telephone calls to explain the situation. The first was to Under Secretary Brown directly. The second was to FEMA's front office. And the third was to FEMA Public Affairs. That third call was to set up a conference call with FEMA Operations in Headquarters, the Emergency Response Team-National team that was based in Baton Rouge, the Regional Response Coordination Center in Denton, Texas, and with FEMA's front office so that I could make as many people aware of the situation that faced FEMA and the City of New Orleans.

Approximately 5 minutes later, I received a call from FEMA Public Affairs asking me how sure I was of what I saw because there was some pushback on the need for a conference call. And I stated I had never been so sure of anything in my life.

In each report and on the eventual conference call, I explained what I saw and then provided my analysis of what I believed to be the most critical issues that we were facing at that moment: Ground transportation into the city was virtually non-existent because of the massive, widespread flooding. Any ground transportation must come from the west of the city because the I–10 bridge to Slidell on the east side of the city was completely destroyed and there was no access from the north because of all the flooded roads and highways.

I also stated that the situation would only worsen throughout the night and into the next day or so because of the massive amounts of water being deposited into the city via the levee break. I described the levee break as being really bad.

Search and rescue missions were critical as thousands of people stood on roofs or balconies in flooded neighborhoods. It was getting dark and the city was without power.

Supplying commodities would be an extreme challenge as more and more people were headed to the Superdome to escape the flood waters, and food and water supplies were already very short at the Superdome. I told them that the Superdome population at that time was now almost 20,000.

Medical care at the Superdome was critical because the staff there had run out of oxygen for special needs patients, and more and more people were in need of medical attention.

Housing an entire city worth of people would be a major issue as approximately 80 percent of the city was under water to varying degrees, and many areas were completely destroyed. And I feared that would increase to 90 or 100 percent with the levee break.

Environmental issues would be a major problem as I reported that an oil tanker had run aground and there was leaking fuel in the river.

I believed at the time, and I still do today, that I was confirming and explaining the worst-case scenario that everyone had always talked about regarding the City of New Orleans.

I then had a meeting with New Orleans Mayor Nagin and his staff of approximately 25 people, and I told them of the situation so that they would know what they faced in the hours and days ahead. I cannot say strongly enough how it was a very emotional meeting for everyone. I could tell they were devastated. Following that meeting, Terry Ebbert pulled me aside and said, “Marty, you have done this before. What do we need to do now?” I told him that he needed to make a list of the city's priorities, actions, and commodities so that they could move forward with an organized plan. I also told him that they needed to let the State know so that FEMA could be tasked to help. He told me, “Consider it done.”

On Tuesday, August 30, I woke up early to discover that water was rising around the Superdome and that by 6 a.m., 2 to 3 feet of water was in the streets around City Hall and the Superdome, and it was continuing to rise by the minute. I immediately called Bill Lokey, the Federal Coordinating Officer in Baton Rouge, and I told him of the rising waters and once again tried to express the seriousness of the situation as I saw it. He thanked me, and he told me he would have an operational team deployed to the Superdome later that day to relieve me.

About an hour later, I was informed that Under Secretary Brown, Governor Blanco, Senators Landrieu and Vitter, and others were planning on flying to the Superdome later that morning. At approximately 8 a.m., I went to find Mayor Nagin, who was at the Hyatt Hotel, to inform him of the visit, and I asked him to meet with the group. So I put on tennis shoes and a pair of shorts, and I walked through what was now waist-deep water over to the Hyatt and told the mayor of the visit. I then walked through the water again to get to the Superdome. I went to the Superdome early for fear that if I waited much longer, I would have to swim.

At the Superdome I spoke with the National Guard to get the latest conditions, and it was obvious that the Superdome conditions were in rapid decline and there was a critical need for food and water. The smell emanating from the Dome was already beyond description. I communicated all of this to Under Secretary Brown when he arrived later that morning. I told him that the Superdome conditions were deplorable and that we desperately needed food and water.

During the subsequent meeting with the mayor and the incoming group, Mayor Nagin pulled out his list of priorities and proceeded to tell everyone what he needed for his city and the residents.

By early Tuesday morning a FEMA medical team arrived at the Superdome, and by early afternoon a four-member Emergency Response Team-Advance (ERT–A) arrived. For the next 3 days, I worked and lived at the Superdome with the ERT–A team and with the FEMA medical team. Each day it was a battle to find enough food and water to get to the Superdome. It was a struggle meal to meal because, as one was served, it was clear to everyone that there was not enough food or water for the next meal. But I can say because of some truly heroic efforts from FEMA staff, the Coast Guard, and the National Guard, enough food was always found and brought to the Superdome, even at the last moment, so that at least two meals were served to everyone each day.

While we battled food and water issues, rescue missions continued, more and more people arrived at the Superdome, and the medical conditions of many at the Superdome were in rapid decline; many people were near death. Working in absolutely deplorable conditions, the New Mexico Disaster Medical Assistance Team saved lives. They worked with helicopter Medivacs to evacuate the most critically ill.

I would like to say that what has been lost in all of the discussions and criticisms of what did or did not happen in New Orleans is that real heroes existed there, and the New Mexico medical team did truly amazing things to save lives and make a difference, as did FEMA's urban search and rescue teams. They worked 24 hours a day to treat patients, and it is accurate to say that without their work, many more people would have certainly died. I am honored that I can call them coworkers because they are truly the best of what FEMA has to offer.

It is well known what happened over the next several days, most of it real, some of it hyped and exaggerated by the media, but all of it tragic. I was and still am today most haunted by what the Superdome became. It was a shelter of last resort that cascaded into a cesspool of human waste and filth. Imagine no toilet facilities for

25,000 confined people for 5 days. Women and children were forced to live outside in 95-degree heat because of the horrid smell and conditions inside. Hallways and corridors were used as toilets, trash was everywhere, and amongst it all, children—thousands of them. It was sad, it was inhumane, it was heart-breaking, and it was so wrong.

By Thursday, plans to evacuate the Superdome were underway. Buses began taking people from the Dome, but there were critical missions to still carry out. But early Thursday, the National Guard approached the FEMA staff and told us that there was intelligence that a riot was being planned for noon and that they did not want to be responsible for our safety, so they recommended that we leave. Phil Parr, who was the senior FEMA official on the ground at that time, made the decision to evacuate all FEMA assets from the Dome, including the medical team. I strongly voiced my concerns about abandoning the mission and the critical need to continue with medical care and the coordination of food and water into the Dome. I pointed out at that time that during the overnight hours, approximately 150 heavily armed forces arrived at the Superdome by helicopter, raising the security level. I called Mike Brown directly to tell him that we were leaving. I contacted FEMA's front office that we were leaving. I think I was hoping that somebody would make us stay. Within an hour, all FEMA personnel were ordered onto trucks and driven out of the Superdome. Our leaving at that moment meant that FEMA had lost all visibility of the situation and operational control at the Superdome. I do not believe that it was the right decision for us to leave.

We were taken to Baton Rouge and the State EOC where I immediately found Under Secretary Brown and again voiced my strong objections about pulling out. He looked at me and he said he was glad I was safe, but I was to "get my ass back there." Within hours, another medical team was dispatched to return to the Superdome, and the next day an operational team was sent back. I never went back. I stayed in Baton Rouge and began to work in public affairs.

My purpose before you today is to help you to understand what happened in New Orleans and when it happened, as I know it from my own perspective. I hope that what you learn from me and the many others that will come later in this Committee is a better understanding of emergency management and response. From this I hope that we are able to effect change so that no other child, no other senior citizen, no other special needs patient, no other parent, and no other community in this country will ever have to experience the horrors and tragedy that happened in New Orleans and the entire Gulf Coast that week.

I will now answer any questions that you may have.

Chairman COLLINS. Thank you very much, Mr. Bahamonde. Your testimony is so eloquent and moving, and even more important, it is a firsthand account. You are an eyewitness to what happened, and that is why your testimony is so important to this Committee.

You stated this morning that it was a struggle meal to meal to find enough food for those who were in the Superdome, and it was only due to the heroic efforts of a lot of FEMA, Coast Guard, and other employees that there was able to be sufficient food for those in the Superdome.

Former FEMA Director Michael Brown stated in his House testimony that you communicated to him that there was "plenty of food" in the Superdome but that later in the week, by Tuesday or Wednesday, they did need additional supplies.

Now, your E-mail to Mr. Brown on Wednesday very clearly states, "We are out of food and running out of water at the Dome."

At any time before this, did you communicate to Michael Brown that there was "plenty of food" in the Superdome?

Mr. BAHAMONDE. I never communicated that to Under Secretary Brown or anyone because I knew that was not the case. As I reported that on Sunday and subsequent E-mails went out, I think that point was clearly made.

Chairman COLLINS. And, indeed, when you spoke to Mr. Brown by telephone on Monday night and again when you met with him on Tuesday morning, didn't you express at that time that there was a critical need for food and water at the Superdome?

Mr. BAHAMONDE. I feel I got the message to him as strongly as I could because I knew on Tuesday when he arrived that there was not any food to feed the people Tuesday evening. So I couldn't have been any more clear to him that food and water was a desperate situation at the Superdome.

Chairman COLLINS. I would now like to turn to the collapse of the levees and the report that you received—that you overheard, actually, at 11 a.m. the day of the storm on Monday in the City of New Orleans Emergency Operations Center.

First, was the transmission clear that this was a break in the levee rather than an overtopping?

Mr. BAHAMONDE. It was absolutely clear. I was standing in their communications room. There were several people around, and I believe I was standing next to Terry Ebbert when that call came in. And they were very descriptive as to where it was, that it was a clean break, because at that part of the levee there is a concrete wall, so there was no overtopping of that wall. They described that the wall had busted through and that water was pouring into the it.

They asked at that time where exactly was the levee break, so we went to a map, and they were able to show me exactly where that levee break took place.

Chairman COLLINS. And you saw the levee break later that day when you went up in the helicopter ride. Is that accurate?

Mr. BAHAMONDE. I did, and it was exactly where they had described it, and it looked exactly as they had pointed it out in that 11 o'clock report.

Chairman COLLINS. The reason I brought up the break versus the overtopping is if water were simply coming over the top of the levee, then the flooding eventually would subside. But if there is a break in the levee, the result is the water is just going to keep flooding into the city. Is that correct?

Mr. BAHAMONDE. That is correct, and that is how it was described in that call.

Chairman COLLINS. So since you heard this report at the city's Emergency Operations Center, is it safe to assume that pretty much the entire city emergency management apparatus became aware of that terrible news?

Mr. BAHAMONDE. Yes, it was clear within the EOC that this news had come in. It filtered out probably over the next half-hour, but it was clear everybody knew what had happened.

Chairman COLLINS. Now, I realize that you were not involved in discussions that may have occurred outside the Emergency Operations Center, although one would think that most of the discussions would be within the Emergency Operations Center on what to do.

You told the Committee that "the city's only plan was to take everybody that they could and get them into the Superdome." I assume that you think that was not a very good plan for reacting to a break in the levee.

Mr. BAHAMONDE. I found it amazing at that time as I watched from the parking lot of the EOC that there were truckloads of 20, 30, or 40 people constantly being sent to the Superdome. I mean, the population of the Superdome almost doubled between Sunday night and Monday night after the hurricane had passed. And you knew the potential that the Superdome was going to flood. You knew what the break was. If you were able to project at all what the information that we had in front of us was going to lead to, it was curious to me that we would then continue to send people to an area that we knew was not sufficient, not supplied, and probably not the best place to collect people, considering that we still at that time had an opportunity to just drive shortly over the bridge, which was really not far from the Superdome, into the Algiers portion of town, which was not flooded, which did not have a levee break, and which was a safe haven for a lot of people.

So I found that very peculiar at that time that that was the plan.

Chairman COLLINS. And, indeed, the decision to keep moving people into the Superdome, which is already under incredible strains and pressure, exacerbated the problems being experienced by the victims who were already in the Superdome. Is that an accurate assessment?

Mr. BAHAMONDE. Absolutely. As I stated already, 40,000 MREs were in place, and for 12,000 people for a couple of days that would last. But when you add to 20,000 to 25,000 people on that, you do the math and you realize you are out of food in a big hurry. And they knew that when they were bringing people to the Superdome, that there wasn't any kind of structure there or commodities to support what was happening.

Chairman COLLINS. Now, there is a critical 16 or even 17 hours between when you heard the transmission at the EOC and, that is, the city was notified of the break in the levee and when the water actually reached the Superdome.

Now, I understand that in the immediate aftermath of a hurricane, conditions are still pretty bad. It is windy. It is difficult. But is there any reason in your judgment why the City of New Orleans, the State of Louisiana, FEMA, all of the emergency management entities could not have done more during that critical 16-hour period to evacuate people not only from other parts of the city but from the Superdome before the flood waters reached the Superdome and made everything so much more difficult?

Mr. BAHAMONDE. In my opinion there was a systematic failure at all levels of government to fully comprehend the magnitude and the severity of the situation that was presenting itself on an hour-by-hour basis. It is the ability to project what is going to happen with the information that you have, and I did not see any action to that effect during Monday afternoon, Monday evening, and even in the overnight hours of Tuesday.

Chairman COLLINS. Now, you mentioned in your testimony that you immediately got on the phone to FEMA officials in the morning after learning of the collapse of the levee to inform them of this very serious new development. You are not sure exactly when Michael Brown was informed in the morning, though there is some E-mail traffic that suggests that it was pretty quick after that. But later in the day, after you went up and saw the break in the levee firsthand, didn't you take a picture, first of all, which we are going to show,[1] and as I understand it, you returned from your second flight and called Michael Brown? Is that correct?

Mr. BAHAMONDE. I did. As soon as I got off that second helicopter flight, really within a couple of minutes, I picked up the phone and I called Under Secretary Brown directly, and I began a 10- to 15-minute conversation that explained everything that I have already explained in my statement.

Chairman COLLINS. What specifically did you tell Michael Brown in that phone call?

Mr. BAHAMONDE. I told him that transportation into the city was non-existent, that all of the roads, highways, bridges were either destroyed or flooded and impassable. And I told him that there was going to be a critical need to get commodities into the city and that you weren't going to be able to do it by ground transportation and that the situation was only going to get worse because the levee that I had witnessed myself, as you can see there,[1] was just pouring water into the city and there was no sign that was going to stop anytime soon.

Chairman COLLINS. And what was his response?

Mr. BAHAMONDE. He didn't ask me any questions. All he said was, "Thank you. I am now going to call the White House."

Chairman COLLINS. Did he indicate what else he was going to do upon learning of this terrible development?

Mr. BAHAMONDE. All he said was, "Thank you. I am going to call the White House."

Chairman COLLINS. Do you happen to know whom he called at the White House?

Mr. BAHAMONDE. I do not.

Chairman COLLINS. I would now like to turn to another issue, and that is the provision of medical care at the Superdome. In his testimony before the House committee, Director Brown said, "Marty both called me on cell phones and E-mailed me on Monday describing the general conditions, you know, that there were 15,000, 20,000 people there, the medical teams were there. It was a fairly routine kind of E-mail."

Did you, in fact, report to Director Brown on Monday that the medical teams were there?

---

[1] Exhibit No. 3 appears in the Appendix on page 62.

Mr. BAHAMONDE. No, because they had not arrived.

Chairman COLLINS. Mr. Brown also testified before the House Committee that FEMA sent a medical team to the Superdome because "that number of people, someone has a cold, someone gets cut, there has got to be something, so we had a medical team there also."

That seems to minimize how dire and serious the medical needs of the evacuees were. Is that assessment correct in any way? Was it just needed because people might have colds or cuts?

Mr. BAHAMONDE. No, that is not correct because on Sunday, while at the EOC, there was an urgent call from the Superdome for oxygen tanks because the special needs patients that were placed there, many were on oxygen, many needed some sort of medical care. That is why they are special needs patients. And there was already a critical shortage, and they were looking to hospitals within New Orleans asking if any hospitals had any oxygen. The hospitals themselves were holding onto their oxygen because they had the same need.

I conveyed that very same information up to FEMA Headquarters, and I know that on that day an E-mail went out to the staff that was supporting Under Secretary Brown that there was a critical medical need in the situation and that oxygen was one of the critical needs, and there was a lot of E-mail traffic to that effect.

So I couldn't make it any more clear at that time to the people that I was talking to at headquarters and at the Regional Response Coordination Center in Denton, Texas, that if they did nothing else, to get a medical team into that Superdome as fast as possible.

Chairman COLLINS. And the medical needs were clearly urgent for some of the evacuees if they were in need of oxygen. But also, later on, on Wednesday, when you personally sent Mr. Brown an E-mail, you talked about dying patients needing to be Medivaced. Estimates are that many will die within hours. So surely Mr. Brown must have understood that the medical team was needed for something far more serious than colds and cuts. Is that accurate?

Mr. BAHAMONDE. It was. It was not a cold-and-cut mission. It was a life-saving mission. And they did that.

Chairman COLLINS. For my final question of this round, I would like to turn to the issue of why you think that your dire warnings were not taken more seriously by FEMA officials. I am trying to understand whether we have organizational problems, leadership problems, resource problems, all of the above.

In my remaining time could you comment on why you think there was not a better, more effective response?

Mr. BAHAMONDE. Well, like I said, I don't think that they were able to comprehend or project what the situation was at that time. I also believe that I was a public affairs person. I wasn't an operational person. I wasn't there to do operations. I was just there as a public affairs person. So I don't know if that carried any less weight with anybody who was listening. But I tried my best to accurately describe what was happening. I tried to do it as it happened. I tried to in the strongest way that I could express what the situation was.

But I truly believe that there was not the inability but they were unable to forecast and project the information that I provided to them as to what could really happen in the next several hours. And that still surprises me.

Chairman COLLINS. Thank you very much for your testimony.

Senator Lieberman.

Senator LIEBERMAN. Thanks, Madam Chairman, and thanks, Mr. Bahamonde. Let me just ask you again for the record how long you have worked for FEMA.

Mr. BAHAMONDE. I have worked for FEMA for 12 years. I started in 1993.

Senator LIEBERMAN. Right, and so you are a civil servant.

Mr. BAHAMONDE. Correct.

Senator LIEBERMAN. Right. And if I heard you correctly, you worked both here in Washington, and more recently you have been Regional Public Affairs Director out of Boston in Region One.

Mr. BAHAMONDE. That is correct.

Senator LIEBERMAN. And beyond that, by your testimony, since 2003, Under Secretary—we know him as FEMA Director—Michael Brown sent you out to do advance work in preparation for a series of natural disasters, including hurricanes, and, in fact, at one point dispatched you to Iran to help with a disaster. Is that correct?

Mr. BAHAMONDE. That is correct.

Senator LIEBERMAN. So I would say from that record that Michael Brown must have had confidence in you, and I presume that is part of the reason why you were dispatched quickly to New Orleans. Correct?

Mr. BAHAMONDE. That is correct.

Senator LIEBERMAN. To the best of your knowledge, were you the only FEMA employee sending reports back from New Orleans during those first days before the hurricane made landfall on Monday, August 29?

Mr. BAHAMONDE. I was the only one because I was the only one there.

Senator LIEBERMAN. You were the only one there. Now, I want to just go into a little more about what you were sending back, and I believe if I refer by number to exhibits, they will put this up on the screen, and you have it there. Exhibit No. 6[1] is an E-mail from you, Sunday, August 28, the day before landfall for Hurricane Katrina. The time is 4:40 p.m., and this is to Michael Heath. Could you identify Mr. Heath?

Mr. BAHAMONDE. Michael Heath was Under Secretary Brown's special assistant in the front office here at headquarters.

Senator LIEBERMAN. OK. And the reason I mention that one is that you say in that E-mail, "Currently 2,000 inside the Superdome with about 300 special needs"—meaning there are 300 people who have some kind of medical conditions, I presume.

Mr. BAHAMONDE. That is correct.

Senator LIEBERMAN. "National Guard is setting up cots. Medical staff at the Dome say they expect to run out of oxygen in about 2 hours and are looking for alternatives."

[1] Exhibit No. 6 appears in the Appendix on page 65.

Now, give me a little background on that. What does that mean? What was the oxygen being used for?

Mr. BAHAMONDE. Well, many of the special needs patients were on oxygen, and so when they were taken to the special needs shelter, there was that requirement that oxygen go with them. But it became clear that this was not going to be a short stay at the Superdome, I think even at that early time. And so they knew that they were going to run out of oxygen, and at that time they were looking at other hospitals within New Orleans. But they all projected that they were going to have that same need when they lost power. And so they were unable to get any additional oxygen into the Superdome.

Senator LIEBERMAN. OK. So I go to this E-mail because what it shows us is that on Sunday afternoon you are sending an E-mail to the special assistant, to Michael Brown making very clear that this is not just people who might get a cold. There are people in there depending on oxygen, and it is about to run out. Correct?

Mr. BAHAMONDE. That is correct.

Senator LIEBERMAN. Do you know whether Michael Heath discussed that E-mail with Michael Brown?

Mr. BAHAMONDE. I do not.

Senator LIEBERMAN. Based on your own experience with Mr. Heath and Mr. Brown in the past, would you assume that he would have conveyed that information to Michael Brown?

Mr. BAHAMONDE. He is a special assistant in the front office. His office is in the main office suite in the front office. So if he doesn't convey it to Under Secretary Brown directly, I would assume that he conveyed it to the staff that was in FEMA Headquarters and that someone would convey that information onward.

Senator LIEBERMAN. OK. Now, I believe you said in your testimony that you took pictures on Sunday and sent them via Internet to FEMA Headquarters. I don't know whether you can call up any of the pictures or a staff assistant over there can, but I——

[Pictures displayed.]

Mr. BAHAMONDE. I think it was the first one.[1]

Senator LIEBERMAN. Yes. And this would have gone to whom? To Mr. Brown's office?

Mr. BAHAMONDE. No. I sent that picture to FEMA Public Affairs.

Senator LIEBERMAN. Ms. Taylor.

Mr. BAHAMONDE. I sent it actually to Bill Koplitz, who was the FEMA photo editor, and I sent it to him so that he could post it on FEMA's website and so that people could get an understanding of what was developing outside——

Senator LIEBERMAN. That is it on the screen right now.

Mr. BAHAMONDE. That is one of the pictures I took, correct.

Senator LIEBERMAN. And why don't you just briefly describe what that is.

Mr. BAHAMONDE. I was standing on an overhead platform looking toward the southern part of the city, and you can see by then that the crowd was already well down to the street into the next block and eventually would wrap around the building. And this was still early on.

[1] Exhibits No. 1 thru 6 appear in the Appendix on pages 60 and 65 respectively.

Senator LIEBERMAN. This is early Sunday?

Mr. BAHAMONDE. Early Sunday, I know that because at approximately 3 o'clock, one of the first feeder bands came through and completely soaked all of the people and their possessions.

Senator LIEBERMAN. Right.

Mr. BAHAMONDE. All the pillows, all the blankets, everything that they brought, were now wet. So the National Guard then took everybody up onto a platform around the Superdome so that they had some protection. But by that time everybody was wet. So I know that picture was taken early even before that happened.

Senator LIEBERMAN. And in response to the pictures, the E-mails about the swelling humanitarian crisis in the Superdome, on early Tuesday morning—it took until early Tuesday morning, I gather, for the first Disaster Medical Assistance Team to arrive, one team. My reaction to that—you tell me; this is your field—is that was, in response to what you were describing, there was a need for more than one Disaster Medical Assistance Team in the Dome. Would you agree?

Mr. BAHAMONDE. Certainly multiple Medical Assistance Teams would have been helpful. I think what happened on that day was there was no anticipation when Sunday started that there would be a need to supply or support the events at the Superdome. But as the city opened it up by around noon and crowds began to arrive, I——

Senator LIEBERMAN. Noon on Sunday?

Mr. BAHAMONDE. On Sunday. I am sorry.

Senator LIEBERMAN. Right.

Mr. BAHAMONDE. I know that FEMA began to discuss getting a medical team in there. The closest medical team at that point was in Houston.

Senator LIEBERMAN. Right.

Mr. BAHAMONDE. Now, medical teams travel in high-profile vehicles, by which I mean large trucks which carry their supplies and everything. And they are really restricted from traveling in winds over 30 to 40 miles an hour. I do know that medical team left Houston late that afternoon and traveled to Camp Beauregard in Louisiana and arrived there at around 11 o'clock that night. But because of the intensifying conditions of the storm, the high winds, the rain, the decision was made that they were unable then to get into the Superdome that night.

I have subsequent E-mails that I think strongly urged in any way possible to get the medical team there because I knew if they didn't get in Sunday, chances are they weren't going to arrive until Tuesday, and we already had a medical crisis in the Superdome.

Senator LIEBERMAN. That leads me to ask this question, and give me a short answer because time is running. As I look at what happened in this disaster and how the Federal Government failed, one of the questions I asked, as I asked in my opening statement, why was it possible for you as a Public Affairs Director to make it into New Orleans in response to the clear warnings that exactly what we feared for years, a hurricane Category 4 or 5 and the flooding of New Orleans, was about to happen and FEMA dispatches one person, the Public Affairs Director, before the storm hits, but not the disaster teams, the search and rescue teams, the emergency response teams. Why did that happen?

Mr. BAHAMONDE. Well, the reason is because one thing FEMA doesn't want to do is to stick staff and resources in the direct path of the hurricane itself, because if you need to move those commodities, you need to move staff, but they are stuck in a situation riding out a storm, they are less able to mobilize and activate. So FEMA takes the approach that we surround the perimeter of where ground zero may be for any kind of a hurricane or disaster, and then when the storm passes, then mobilize their assets into that location.

Senator LIEBERMAN. But I must say that is something I think we all have to, as this goes on, ask ourselves whether it makes sense. You were able to get in. Wouldn't we all have been better off, wouldn't the people of New Orleans have been better off if the search and rescue, medical assistance, and emergency response teams had also gotten in? I mean, that is where disaster response people ought to be. That is where the local emergency operations people were, which is where the storm was most likely to have the most damage.

I want to come back to something else, another question I raised. You testified—Senator Collins asked you questions about, from early Monday morning you had received information that the levees had broken, Monday, August 29.

Mr. BAHAMONDE. That is correct.

Senator LIEBERMAN. The storm hits New Orleans, and the waters begin to surge. You sent E-mails—there is validation of that—about 1:40 and an E-mail from Cindy Taylor that we have as part of the record.[1] During the day you have the conversations that you have described. You even take pictures—which maybe while I am talking you could put up pictures—of the flooding on Monday back to FEMA Headquarters. And yet I want to read to you comments from Secretary Chertoff on "Meet the Press" on the following Sunday, where he says, "It was on Tuesday that the levee"—"it may have been overnight Monday to Tuesday that the levees started to break, and it was midday Tuesday that I became aware of the fact that there was no possibility of plugging the gap and that essentially the lake was going to start to drain in the city."

Secretary Rumsfeld, on September 6, in a briefing at the Pentagon, says, "The original blow was the storm. New Orleans escaped a great portion of it. If you are talking about New Orleans." The flood followed that by a day, Tuesday.

General Myers said, "I think before the storm even hit, actions were being undertaken in the Pentagon." He says, "On Tuesday, at the direction of the Secretary and the Deputy Secretary, on Tuesday we went to each of the services. I called each of the chiefs of the services one by one and said, 'We don't know what we are going to be asked for yet.'"

On Tuesday, after you have been saying for more than 24 hours that the levees have already broken, the Weather Service has been saying for more than a week that this is going to be the feared Category 4 or 5 hurricane. On Tuesday, General Myers says, "We don't

[1] Exhibit No. 14 appears in the Appendix on page 66.

know what we are going to be asked for yet. The levees and the flood walls had just broken, Tuesday."

Now, how do you explain that after you, FEMA's person on location, were saying from Monday morning—talk about precious hours. Senator Collins talked about those 16 hours. The floods are beginning to rise. When the 17th Street levee broke, it took a while. The 9th Ward, when the Industrial Canal broke, flooded almost immediately. The Lake Pontchartrain flooding, as the lake came into the City of New Orleans, took a day and a half.

I just say to myself, if they had listened to you early and gotten more emergency teams in, more Department of Defense personnel, maybe some lives that were lost could have been saved. How do you explain those comments about Tuesday from Secretaries Chertoff and Rumsfeld and General Myers?

Mr. BAHAMONDE. I don't know if I have an explanation. I know I was there to provide information. I gave it when I knew it. I gave it to as many people as I possibly could. Maybe I wish I would have called them instead. It is easy to Monday morning quarterback, but I knew at 11 o'clock, FEMA Headquarters knew at 11 o'clock, Mike Brown knew at 7 o'clock, most of FEMA's operational staff knew by 9 o'clock that evening. And I don't know where that information went.

Senator LIEBERMAN. My impression from some of the testimony you gave to our staff was in that 9 o'clock conference call with FEMA staff—and you correct me if I have misinterpreted this—you felt there wasn't the level of response that you hoped for, that the response seemed to almost be that, well, you are not telling us anything we don't know. But it seemed to me that you were telling them something that at least Secretary Chertoff and Secretary Rumsfeld apparently did not know until the next morning.

Mr. BAHAMONDE. When I made that conference call, I remember when the conference call ended, Scott Wells, who is the Deputy Federal Coordinating Officer, said to me on the call, he said, "Thanks, Marty. We already knew most of this information, but thanks for providing it." And that comment struck me and it stays with me today because I know when I went up in that helicopter, there were only three helicopters in the air at the time, all Coast Guard. And when we flew over the bridge to Slidell, the Coast Guard helicopter said to me, "Wow, nobody has been out here before. I wonder what we are going to find."

So I knew that there were maybe five people in the whole world, those of us that were on that helicopter, that knew that information. And so I was disturbed when somebody said, "We already knew all that," because I knew it just wasn't possible.

Senator LIEBERMAN. Mr. Bahamonde, is the picture on the screen now one of those you took from the helicopter on Monday afternoon?

Mr. BAHAMONDE. That is. That is a picture of a neighborhood,[1] and the reason I took that picture was, if you can see how the buildings are structured, you can already tell that homes were already taken off their foundations. They are all mixed up together. And it already told me that the water depth in many places was

---

[1] Exhibit No. 4 appears in the Appendix on page 63.

10, 11, 12 feet deep, and that was, you know, at 5 o'clock Monday afternoon.

Senator LIEBERMAN. Do you remember whether that was the 9th Ward or——

Mr. BAHAMONDE. No. This picture was taken out closer to where the 17th Street Canal levee break was.

Senator LIEBERMAN. So this is the beginning of Lake Pontchartrain draining into New Orleans.

Mr. BAHAMONDE. Absolutely.

Senator LIEBERMAN. And it looks from here like the water is up in some cases close to the roof of the dwellings.

Mr. BAHAMONDE. I think many of the homes that you can see there, all you see is the rooftop.

Senator LIEBERMAN. Yes. And, again, when did you send that to FEMA Headquarters in Washington?

Mr. BAHAMONDE. I actually did not send this photo because at that time there was no power, there was no Internet service, there was no capability of doing it. I took it just for the record.

Senator LIEBERMAN. But this is the scene you described in the direct conversation with Michael Brown that evening and in the conference call at 9 o'clock.

Mr. BAHAMONDE. Absolutely.

Senator LIEBERMAN. Thank you.

Chairman COLLINS. Thank you. Senator Warner.

OPENING STATEMENT OF SENATOR WARNER

Senator WARNER. Thank you, Madam Chairman.

As I sit and listen to this testimony, I reflect on the history of military organizations, and I have spent much of my life studying it, and with some limited extent myself in situations. The enemy here was Mother Nature, but many times have military units been overcome with situations where those in command and those trained to do what had to be done just are not able to do it, and the responsibility falls to someone who may not have the military occupational specialty to do it, but they do it anyway. And apparently, I think, from what I gather here in this brief few moments of the testimony, you discharged yourself very well and very commendably, drawing on your assets and experience and not just limiting yourself to that of a public affairs officer.

In that historic call to Brown where you described things and so forth, did you suggest that there be persons of operational qualifications dispatched immediately to come down there?

Mr. BAHAMONDE. Well, I think it was obvious that the need existed to provide some coordination to what was going to be needed at the Superdome, yes.

Senator WARNER. But you didn't specifically say to him, "Hey, boss, we got a problem here, and you better send some of the trained operators down as fast as possible" or something like that?

Mr. BAHAMONDE. No, I didn't specifically request that. I assumed, the way that we always do business, that would happen.

Senator WARNER. Was there any protocol problem, namely, with the timing of the request from either the governor or the mayor or something that should trigger the entrance of FEMA people?

Mr. BAHAMONDE. Well, like I said, I think FEMA always has taken the position not to put people in harm's way. And I think what surprised me is that there was capability to get a team in there on Monday but that the team didn't arrive until Tuesday around noon.

Senator WARNER. You said very forthrightly there was a systematic failure at all levels of government to fully recognize the seriousness of the situation. Were the facts available in your judgment to those in our government at different levels to make an assessment of the seriousness of this situation?

Mr. BAHAMONDE. I believe so, yes.

Senator WARNER. You have had other experience, and you have been the man on the scene many times. Was this in your judgment the seriousness that came to your professional background as one that demanded prompt attention? There was no hesitation on your part to make an assessment of the facts and the needs and the requirements that should be brought to bear, was there?

Mr. BAHAMONDE. In all my years in FEMA, there has always been a discussion of what the worst-case scenario would be, and it always centered around a Category 4, Category 5 hitting New Orleans. And like I said, I have already testified, I think at that time I was describing that worst-case scenario because it was happening, it was developing. And so I don't think anybody should have been caught short that there wasn't a serious situation developing.

Senator WARNER. Now, you described the three helicopters, and they were all Coast Guard, but I think the record should reflect you saw no helicopters provided by the City of New Orleans or the State functioning at that point in time, did you?

Mr. BAHAMONDE. I did not see any. That doesn't——

Senator WARNER. Do you have any information that others may have been operating?

Mr. BAHAMONDE. I do not. All I saw was Coast Guard helicopters.

Senator WARNER. Well, presumably had there been others up, they would have been coordinating with the Coast Guard, if only for personal safety flying around in the same airspace.

Mr. BAHAMONDE. And I do know that when we were doing the flyover, that was one of the concerns for the Coast Guard. We were identifying where other Coast Guard helicopters were because the last thing they wanted to do was duplicate rescue efforts at the same location.

Senator WARNER. Were the city or State emergency operations or first responders positioned to report regarding infrastructure damage, rescue needs, or the situation at the Superdome?

Mr. BAHAMONDE. Well, there was somebody that was situated out by that levee break because they had an eyewitness account. But I don't know the extent of where the city had placed people in and around the city to provide situational awareness.

Senator WARNER. Well, I thank you very much. I thank you, Madam Chairman. This is a very important hearing in your record on this subject.

Chairman COLLINS. Thank you. Senator Levin.

OPENING STATEMENT OF SENATOR LEVIN

Senator LEVIN. Thank you, Madam Chairman and Senator Lieberman, for your thorough commitment to oversight, which is so essential. This Committee is, again, leading the way, and it is important that we do so, and I commend you both on it.

And I want to commend you, Mr. Bahamonde. Give me the official pronunciation. I think our Chairman and the others have pronounced it correctly, but let me make sure. Is it Bahamonde?

Mr. BAHAMONDE. Bahamonde.

Senator LEVIN. I want to thank you for your direct testimony, for your courage in what you did in New Orleans. The story you tell is incredibly discouraging, incredibly powerful, and incredibly straight. And it conflicts in very serious ways with the story that the Director, Mr. Brown, told the House of Representatives, and I want to go into those differences in a moment because there are some very serious conflicts between you and the Director, the FEMA Director.

But before I do that, I want to just pick up on one thing that Senator Lieberman mentioned, and that has to do with the fact that Secretary Chertoff, Secretary Rumsfeld, and General Myers—these are the three people who are in charge really of implementing a response, one head of the Defense Department, a civilian; the other one, head of our military; and the other one, the head of the Department of Homeland Security—say that they didn't know about the break in the levee and the flooding that resulted until almost a day after it occurred, and as the Chairman points out, 16 hours or more after you directly told Mr. Brown, the FEMA Director. That is incredible. It is a huge breakdown of some kind in communications.

But at least one thing they might have learned is if they had watched television on Monday night because Mr. Brown was on television Monday night, and he said that—and apparently this is right after you spoke to him, but at 9 o'clock, for instance, on MSNBC he says, "I have already told the President tonight that we can anticipate a housing need of at least tens of thousands," and he reported on television that his folks in the field—I guess you are it—have reported to him that literally tens of square miles are inundated with water up to the roofs.

So your message got through to Brown. His message was on national television of the flooding, the massive flooding, tens of square miles of homes. And that message apparently got to the President from Mr. Brown on Monday night, but yet we have our three implementing leaders incredibly saying that they didn't know about anything this significant until 16 to 24 hours later.

And I know the Committee will be inquiring of them as to how that is possible. Where was that breakdown? It is not just in communicating one to one or through the usual channels of communications. It is on television. On CNN at 10 p.m. on Monday, Mr. Brown says, "We now have literally neighborhood after neighborhood that is totally engulfed in water." So it was clear on Monday night to Mr. Brown and at least to the rest of the world—and to the President, apparently—that we had this kind of a massive problem. And yet Chertoff, Rumsfeld, and the Chairman of the Joint Chiefs didn't learn of it, they say, until the next day.

That is something which I know the Committee will be pursuing, and you cannot answer, but I just wanted to add that one element that there were national television stories covering Mr. Brown's statements, which obviously followed your conversation with Mr. Brown since there was no other source that he would have.

The question I want to talk to you about and I need your responses has to do with the differences in what you have testified to and what Mr. Brown told the House of Representatives. How many people, FEMA people, were prepositioned at the Superdome or in New Orleans?

Mr. BAHAMONDE. One.

Senator LEVIN. And that was?

Mr. BAHAMONDE. Me.

Senator LEVIN. OK. Now, you have Mr. Brown telling the House of Representatives something very different, and maybe you can explain it. Maybe there is an explanation for it, but it seems a direct conflict to me. This is Representative Thornberry saying that he understood that the Emergency Coordinator for the city had chosen the Superdome as a shelter for people in the city, and that Mr. Brown said he did not make that decision, that was the city's decision. And Thornberry said, "It was something the city chose. You knew about it." "That is correct." And then Thornberry says, "So you have testified that you put some water and MREs there and some people before the hurricane hit." "Put some people in the Superdome before the hurricane hit."

"What were the people supposed to do?" Representative Thornberry said, and here is what Mr. Brown says: "There were two groups of people. One was a FEMA team, kind of a mini-Emergency Response Team to help, again, push through the standard unified command structure, the needs of what was going to be in the Superdome."

Do you know, was there any FEMA team, any mini-Emergency Response Team there in the Superdome to do that?

Mr. BAHAMONDE. Not prior to the storm.

Senator LEVIN. I mean prior to the storm.

Mr. BAHAMONDE. No.

Senator LEVIN. Because that is what he is talking about. And then he said, "It consisted of a gentleman by the name of Phil Parr, who is one of the Federal Coordinating Officers, a very smart man, and a career employee by the name of Marty Bahamonde, who has been with FEMA for numerous years, an excellent employee. I put those two in to help liaison between the mayor's office and the State EOC and myself."

I mean, you were there, obviously. Was Mr. Parr there before the—was he prepositioned there?

Mr. BAHAMONDE. No, he wasn't.

Senator LEVIN. And then he said, "Then we also put a contingent of the National Disaster Medical Team because we knew there would be medical needs for people, you know, a number of people." And this is what the Chairman has made reference to, that someone has got a cold, someone gets a cut.

"So we had a medical team there also." And then Representative Thornberry says, "And so how many total FEMA people were prepositioned, approximately, at the Superdome?" Mr. Brown:

"Counting the team, which I will count as FEMA people, you know, a dozen."

Prepositioned at the Superdome. Were there a dozen people prepositioned by FEMA at the Superdome counting those teams?

Mr. BAHAMONDE. No, there were not.

Senator LEVIN. All right. Well, Madam Chairman, I think you also pointed out a conflict in some of the statements of Mr. Brown. The testimony that you elicited had to do with the MREs and the water, I believe, where there are just clear conflicts here between the testimony which we have received this morning and the House of Representatives. And obviously it is up to the Chairman, but I think it would be appropriate that the House be informed that there are some serious conflicts. Mr. Brown testified under oath. As I understand it, he testified under oath.

And so I would expect that they will read about this in the paper, but more than that, it seems to me that this is—and these are not the only conflicts—that this is significant enough that our staffs need to pull out the conflicts from the testimony, or the apparent conflicts from the testimony, and to refer them to the House of Representatives.

Chairman COLLINS. Senator, I would say that I know the House is following our hearings very closely. I, too, in my opening statement made reference to the issue that you just raised about the prepositioning of personnel. There is a discrepancy on food. There is a discrepancy on a variety of very critical issues. And obviously one of the important things we are doing here today is establishing a record that we will subsequently ask Mr. Brown about when he is called before the Committee to testify.

Senator LEVIN. All right. I thank the Chairman.

Another area that I would like to go into with you has to do with the time when you left. You were pulled out of there basically by the National Guard, as I understand it. They told you that you should leave, that you were not safe. You disagreed with that decision very strongly, and you left nonetheless. I mean, you were told to leave, and you followed the orders. And, by the way, I have no quarrel with what you did. Quite the opposite. You stood your ground as long as you could, and the National Guard Director said you folks have got to leave. So my issue is not that at all. I commend you for the effort. You put up an argument, you put up the struggle that you put up to stay there. It was typical of the kind of courage that you showed.

But I want to go beyond that as to what followed from that because there were apparently buses that had been lined up to remove people from the Superdome first and then from the Convention Center. Those buses were sitting at some kind of a restaurant, I believe, 25 miles outside of New Orleans, waiting for directions to come and evacuate people from the Superdome.

When you and your other folks by then, your teams, were told to leave, something very critical fell through the cracks, and those were those hundreds of buses that were sitting there waiting for directions to come and evacuate people first from the Superdome.

So far am I accurate? Because it is a long question. Is that basically what happened, that there were those buses that were lined up on Wednesday and Thursday waiting for the direction to come in and evacuate?

Mr. BAHAMONDE. Well, the buses were already deployed and at the Superdome when we were evacuating because as we left the Superdome, we were passed by several buses with evacuees on them. So I know that effort had already begun.

Senator LEVIN. All right. There was a newspaper article that said there were hundreds of buses that were waiting, I think, at a truck stop—let me just quote the *Dallas Morning News.* It made reference to a Texaco truck stop on the interstate near La Place, Louisiana, 25 miles outside of the city. Are you familiar with that? Were you involved in trying to get buses to locate there, to deploy there?

Mr. BAHAMONDE. No, I didn't have any knowledge of where those buses were at that time.

Senator LEVIN. OK. Is the name of the National Guard official in Louisiana General Veillon? Did you have any contact with him?

Mr. BAHAMONDE. Not specifically. That doesn't ring a bell.

Senator LEVIN. All right. Let me read this to you, and I will read you this one paragraph and tell me whether you have any knowledge about this from that article. "He said that FEMA's departure left Guard officials scrambling to keep the bus evacuation moving. He said he remembered someone from FEMA mentioning a Texaco truck stop on the interstate near La Place, 25 miles outside of the city, would be a staging area for incoming bus convoys. He"—the general—"flew there the next morning"—which I guess would be Friday morning—"and found several hundred buses. Some of them had been waiting for orders for a day."

And then he the next morning, now Friday, sent food to the Convention Center and on Saturday began moving people out from the Convention Center.

Here is the issue: Because of that day delay, according to this article, if it is accurate, the people were delayed a day, first being—a lot of people were not evacuated that could have been evacuated, first from the Dome, and then a day later than they would have been evacuated from the Convention Center because of this mess-up, confusion.

Do you know who it might have been from FEMA who mentioned those hundred buses waiting at a truck stop to the Guard as FEMA people were told to leave? Do you know who that would have been?

Mr. BAHAMONDE. Well, the person in control of the operation for FEMA at the Superdome was Phil Parr, and I know he was the one that was most instrumental in dealing with the bus issue. So I would assume if that conversation took place, it took place with Phil Parr.

Senator LEVIN. All right. But then you weren't privy to that conversation?

Mr. BAHAMONDE. No, I wasn't.

Senator LEVIN. And just to be real clear, again, on the testimony, the medical team which arrived at the Superdome was not prepositioned, nor was any team prepositioned at the Superdome prior to the hurricane.

Mr. BAHAMONDE. You are correct.

Senator LEVIN. Thank you. Thank you, Madam Chairman.

Chairman COLLINS. Thank you. Senator Dayton.

OPENING STATEMENT OF SENATOR DAYTON

Senator DAYTON. Thank you, Madam Chairman, for holding this very important hearing.

Mr. Bahamonde, I commend you, as others have, for your courage, both in being in New Orleans during that critical time and also coming before this Committee.

I want to refer also to testimony that Mr. Brown made, and he did indicate that he knew he was under oath just before he referenced the—or made some of the comments that Senator Levin just referenced. But Mr. Brown then went on to say, regarding this so-called team that was in New Orleans prepositioned, and he did indicate you in particular. He said, "In fact, Marty both called me on cell phones and E-mailed me on Monday describing the general conditions. You know, there were 15,000, 20,000 people there. The medical teams were there. It was a fairly routine kind of E-mail."

He then goes on to say, "Then Marty later was able to communicate to me the information that they had plenty of food, but by Tuesday or Wednesday, they did need additional supplies, and he was trying to get additional supplies there."

I then go to your E-mails. According to the records we have, on Sunday evening you are E-mailing Michael Heath saying, "Medical staff at the Dome say they expect to run out of oxygen in about 2 hours." Again, Sunday evening, "Everyone is soaked. This is going to get ugly real fast. Everyone here at the EOC is very concerned with what might happen here."

Again, Sunday evening, "Our intel is that neither the Oklahoma DMAT nor the public health officers staged in Memphis will make it to the Superdome tonight. Oxygen supply issue has not been solved yet either. But I talked to the front office and was told that this was the 'solution.'"

It seems to me there is great disparity between what your E-mails—and I don't know where in the chain of authority with FEMA they were directed, but the Director is saying that as of Monday he is getting routine E-mails from you that the situation is seemingly reasonably well in hand, and your E-mails that preceded that Sunday evening seem to indicate to me that you were aware this was a very critical situation and, as you said, getting ugly real fast.

Can you explain that disparity?

Mr. BAHAMONDE. I can't other than I know that nothing I did was routine as I tried to express in the best way that I could, and maybe I didn't express it well enough, the urgency and the need for a medical team before the hurricane hit because there was already a critical situation developing there on Sunday.

Senator DAYTON. Would you expect, knowing how FEMA operates, that the Director would be knowledgeable at the time on Sunday or at least on Monday of how many FEMA employees actually were on site in New Orleans?

Mr. BAHAMONDE. I would expect that he would know.

Senator DAYTON. So when he would testify subsequently, even having had a chance to refresh his memory as to a dozen, including a medical team that were prepositioned, when, in fact, you were the only one there, that would be information that you would expect him to have?

Mr. BAHAMONDE. I would hope so, yes.

Senator DAYTON. Would you expect these kinds of communications that you made on Sunday evening regarding the conditions in New Orleans and at the Superdome, would you expect that to have been transmitted by these individuals—Michael Heath, Cindy Taylor, and Deborah Wing? Would you expect that to have been communicated to the Director?

Mr. BAHAMONDE. Well, I would certainly hope that contacting Mike Heath—and that is one of the reasons why most of my conversations, most of my phone calls and E-mails were to Mike Heath, because I knew he was the special assistant to the Under Secretary and was in the front office, and if it wasn't him directly, that it would have been the chief of staff or the deputy or somebody would convey that information.

The other information that went through Cindy Taylor I know was passed on to Natalie Rule and Nicol Andrews. She's the Director of Communications for the agency and the Deputy Director of Communications. At some point, you think enough people know that the word would have to get to the Under Secretary.

Senator DAYTON. In your testimony this morning, you say that the National Guard told you that they expected 360,000 MREs and 15 trucks of water to arrive that Sunday evening. As the storm intensified outside, instead of those numbers, only 40,000 MREs arrived; instead of 15 trucks of water, only 5 arrived; and the medical team did not arrive either.

You have described why the medical team couldn't make it. What was the reason that so many of the MREs and the water did not arrive?

Mr. BAHAMONDE. I don't have any idea.

Senator DAYTON. When you said the National Guard told you that they expected 360,000 MREs—this is on Sunday evening—did that seem like an appropriate response at that time, a necessary number of MREs to have to handle the situation?

Mr. BAHAMONDE. Well, it certainly sounded like good news. I mean, 360,000 MREs for 12,000 people would tell me that they could stay there for quite a while.

Senator DAYTON. Right. And could you indicate the chronology then? Was food and water then forthcoming during the next couple of days? I think you indicated that you were able to serve two meals a day and that there was sort of a just-in-time kind of flow of food and water. Is that correct or not?

Mr. BAHAMONDE. The difficulty with the food at the time, once we knew we didn't have any, was trying to locate it within the arena of New Orleans and how to get it in. I specifically remember that I worked diligently with the person who was commanding the Coast Guard helicopters, and I specifically asked them, Can you send a helicopter out to a location where I had been told there were five trucks of food and water—but I had no idea if they were actually there. There was just a report that they were. And I wanted the Coast Guard to fly out there to actually see and let me know that is where some food was. The Coast Guard did fly out there, confirmed it, landed, cut open the trucks—because there were no drivers left with those trucks. They were just dropped at a location—and started to put food and water into their helicopter and deliver it to the Superdome. If that didn't happen, there was not going to be any food there Tuesday night.

Senator DAYTON. This is an E-mail from Cindy Taylor to you, and I guess actually it was an original message from Sharon Worthy to a Valerie Smith, so I will correct that: "Please schedule Joe Scarborough this evening for the 9 p.m. Central Standard Time period. I spoke with his producer and told him to call you. Mr. Brown wants to do this one. Also, it is very important that time is allowed for Mr. Brown to eat dinner. Given that Baton Rouge is back to normal, restaurants are getting busy. He needs much more than 20 or 30 minutes. We now have traffic to encounter to get to and from the location of his choice, followed by wait service from the restaurant staff, eating, etc. Thank you, Sharon Worthy, Press Secretary."

Is this a typical command mode while a crisis of this magnitude is unfolding?

Mr. BAHAMONDE. It hasn't been my experience.

Senator DAYTON. At this point, the Under Secretary was in Baton Rouge?

Mr. BAHAMONDE. Correct.

Senator DAYTON. And is that the appropriate location for him given the communications network, the transportation difficulties and the like, or should he have been elsewhere, positioned elsewhere?

Mr. BAHAMONDE. Well, I think it was appropriate that he was in Baton Rouge. I mean, that is where the EOC, the State EOC, that is where really everything that was happening, the command center, so to speak, was coming out of the EOC. So I found it appropriate that he was in Baton Rouge.

Senator DAYTON. Who made the decision, I guess on Sunday, to house people who could not evacuate at the Superdome?

Mr. BAHAMONDE. I believe that was the mayor.

Senator DAYTON. And once the doors opened, whoever could get inside just entered?

Mr. BAHAMONDE. Once it was made known that the Superdome was now a shelter of last resort, people came in droves. And it struck me because people were pulling up in pickup trucks and literally unloading the back of pickup trucks with coolers and boxes and pillows, and my impression was, Did they expect to stay for a couple of weeks? Because I was floored by how much stuff people were bringing to the Superdome on Sunday.

Senator DAYTON. And then when the levees broke and the flooding began, that drove more people to the Superdome?

Mr. BAHAMONDE. Correct, and thousands more were either walking to the Superdome, they were being driven in National Guard trucks. There were Fish and Wildlife trucks that were bringing people. It was a constant flow of people really starting mid to late Monday afternoon, the rest of that evening.

Senator DAYTON. You describe wading in water up to your waist—is this correct?—to the Hyatt, which is—is that across the street from the Superdome? What was the level of the water at its peak relative to the entrances to the Superdome?

Mr. BAHAMONDE. I think there is a picture of the situation right outside the Superdome where you can tell the level of the water depth by cars.[1] I mean, there were cars that were literally underwater.

Senator DAYTON. What prevented the water from flooding the Superdome? The doors were all closed at that point?

Mr. BAHAMONDE. Well, the Superdome is elevated. It is probably elevated about 15 to 20 feet above ground level to where people access it. Certainly the basement of the Superdome was flooding because I took a picture of that as well, and that is a picture of right outside the Superdome. You can see some cars to the left there were almost up to their roofs, and other cars. So it was a good four feet.

Senator DAYTON. With the benefit of hindsight, but at that time when the local authorities made the decision to use the Superdome as a refuge of last resort, was that the proper decision at that time?

Mr. BAHAMONDE. Well, in hindsight, no.

Senator DAYTON. What would have been the better alternative?

Mr. BAHAMONDE. What surprised me is when—I don't think I had a problem with them putting people in the Superdome before the storm because there aren't many places where you can stick 12,000 people and have them safe. What surprised me was after the storm, when the flood waters were as they were, that thousands more people were brought to the Superdome because, A, you knew you were not going to be able to get commodities in there very easily; B, you knew there were no commodities in there; and so you just doubled the population and really took a bad situation and made it so much worse when you had options to just drive them out of the city across the bridge. You were driving them into the Superdome. Why didn't you just take them across the bridge?

Senator DAYTON. This was before or after the—this is after the levee broke and——

Mr. BAHAMONDE. This was after the levee broke. This was, you know, on Monday afternoon.

Senator DAYTON. And it was still at that point possible to transport people on the roads, the highways to get them out at that point?

Mr. BAHAMONDE. Right, because Monday afternoon there was no flooding around the Superdome. This picture was taken on Tuesday morning, but even Tuesday morning they continued to bring people to the Superdome.

Senator DAYTON. And did anybody discuss at this point the fact that they were—so the city officials were actively—somebody was actively driving people to the Superdome. Was there any discussion that they were overloading the Superdome and that they ought to go elsewhere?

Mr. BAHAMONDE. I wasn't privy to any of that conversation.

Senator DAYTON. Who was having those conversations? You have the Federal, State, and local authorities all with some shared responsibility for this.

Mr. BAHAMONDE. Well, you have to understand, at that time all of those decisions were being made by the City of New Orleans and

---

[1] Exhibit No. 5 appears in the Appendix on page 64.

I am assuming the State because it was the National Guard that was helping to bring people into the Superdome. They were actually providing the transportation. I can only assume that the city was telling them where to take them. But FEMA at that point had no decisionmaking responsibility as to where to take these people or where to put them. We weren't in that game at that time.

Senator DAYTON. Is that the right locus for that kind of decisionmaking authority under those conditions, to have the local officials making those kinds of decisions?

Mr. BAHAMONDE. It always is. The local government, State government is responsible for what happens in their cities prior to a hurricane, whether it be evacuation or shelters, absolutely.

Senator DAYTON. And post-hurricane, with the city flooded, that still is properly the local decision?

Mr. BAHAMONDE. All disasters start locally.

Senator DAYTON. And then FEMA's role is to come in and whatever that decision is, try to make that decision work, carry it out?

Mr. BAHAMONDE. We operate under the recommendations that come from the city and the State as to what they need, what they want, and what their wishes are.

Senator DAYTON. So if a seemingly wrong decision is made or is underway, is it FEMA's role somewhere along the line to indicate that they think with their expertise that this is a mistake or is leading to an undesirable outcome? Or, again, is FEMA's role just to carry out and try to make the best of whatever situation?

Mr. BAHAMONDE. No, I think we actually try to influence them on decisions. I mean, we are the ones that do this all the time.

Senator DAYTON. Right.

Mr. BAHAMONDE. We do it 50, 60, or 70 times a year, so I think we do try to provide some management into what they are going to do. But you have to understand, the States and the cities have their plan, and I think the last thing that we want to do is come in there and interrupt their plan because that is what they have trained for. But we do try to advise them on what to do, yes.

Senator DAYTON. Thank you, Madam Chairman.

Chairman COLLINS. Thank you. Senator Akaka.

OPENING STATEMENT OF SENATOR AKAKA

Senator AKAKA. Thank you very much, Madam Chairman. I want to add my welcome to Mr. Bahamonde. Hafa Adai.

Mr. BAHAMONDE. Hafa Adai. I used to live on Guam, sir, so I know what you are talking about.

Senator AKAKA. I appreciate the openness and candor you have displayed while working with our Committee over the past few weeks. Before I get into my line of questions, I want to ask a couple of questions to clarify some of what you have just testified to.

When you briefed FEMA Headquarters on the levee breach at 11 a.m. Monday—I understood you were there before landfall of the hurricane, which was on Sunday, and you witnessed these. But at 11 a.m. on Monday morning, there was a breach report. Eight hours later, you briefed Under Secretary Brown on the levee breach.

Did it seem that Under Secretary Brown knew about the levee breach from his staff by the time you briefed him or was it the first time he heard about the broken levees?

Mr. BAHAMONDE. I was unable to tell because it really was a one-way conversation. I called him, and I did all the talking. He didn't ask any questions. All he stated to me when we were done was, "Thanks. I need to call the White House." So I was unable to determine whether he knew or did not know before that.

Senator AKAKA. I understand that you were the first FEMA person on the scene there. When did the next FEMA person arrive in the area?

Mr. BAHAMONDE. The next FEMA asset to arrive was the medical team, which arrived at approximately 3 a.m. on Tuesday. Under Secretary Brown arrived for his meeting at approximately 11 o'clock that morning on Tuesday, and the operational team that was led by Phil Parr of the ERT team didn't arrive until about noon. So Under Secretary Brown was actually there before the operational team arrived.

Senator AKAKA. My understanding of FEMA is that FEMA is a coordinating agency, as you pointed out, and they work with the local and State governments and even with other Federal agencies, but primarily as a coordinating agency.

You spoke much about the Superdome and how atrocious the situation became. Who was in charge at the Superdome?

Mr. BAHAMONDE. It was the National Guard.

Senator AKAKA. Also, did you notice if the Army Corps of Engineers was there?

Mr. BAHAMONDE. I didn't. I didn't have any conversations with them. It was primarily the National Guard and the mayor.

Senator AKAKA. In your interview with the Committee staff, you stated that once you were deployed to New Orleans, it was up to you to find a local Emergency Operations Center, EOC, in which to ride out the storm. And as you mentioned, I think that was set up in Baton Rouge.

Mr. BAHAMONDE. No. I set that up.

Senator AKAKA. You did. If I understood your statement correctly, FEMA Headquarters gave you a list of EOCs to contact once you arrived in New Orleans and left the choice up to you. Is that correct?

Mr. BAHAMONDE. That is correct.

Senator AKAKA. According to your statement, you took shelter in the New Orleans Parish EOC, which happened to be located across the street from the Superdome. It is my understanding from your statement that you served as a FEMA liaison to Mayor Nagin because you were in the right place at the right time. However, you were not specifically tasked with that role. Is that correct?

Mr. BAHAMONDE. No, I was not. When I was sent to—the directions that I received from FEMA Headquarters were to go to New Orleans, not go to any specific place, not specifically talk to anyone. It was just to go to New Orleans.

Now, my experience in having done this before, I contacted David Passey, who is the regional external affairs director in Region Six in Texas, and I said you have got to give me some names of people I can call where I can ride out the hurricane. And my experience is you always ride out a hurricane at an Emergency Operations Center. They are the ones that are designed to be able to sustain that.

He gave me a list, and I actually called Jefferson Parish first, and they didn't answer their phone. And so my next call was to Orleans Parish, and they answered the phone, and I spoke to a woman named Sariah, who I think was the deputy director of the EOC. We had a short conversation, and she welcomed me to come to the EOC. And so I went there with the primary mission to just ride out the storm, be in a place where I could gather some information, but I was not sent there to be a liaison with anyone.

Senator AKAKA. Would it be fair to say that former Under Secretary Brown's statement that he sent a FEMA representative specifically to Mayor Nagin's office prior to landfall was incorrect?

Mr. BAHAMONDE. That is true. That is incorrect.

Senator AKAKA. You have stated that you felt your presence at the Superdome was critical because FEMA was the sole source of food and medical attention for the evacuees, and that you strongly objected to the evacuation, as you did say, of FEMA staff.

When you told Under Secretary Brown that FEMA staff were evacuating the Superdome, did you voice your concerns to him? If so, what was his reaction? Did Under Secretary Brown agree that FEMA would lose its visibility of the situation and operational control at the Superdome?

Mr. BAHAMONDE. No, I didn't specifically tell him my opinion that we should stay. I called him, and I said, "Sir, I want you to know that a decision has been made to evacuate us from the Superdome because the National Guard has just told us that they no longer want to be responsible for our safety, so they recommend that we leave." And I said, "And so a decision has been made by Phil Parr that we have to get out now." And his comment was, "Well, be safe," and that was the gist of the conversation.

Senator AKAKA. When you met with Secretary Chertoff in Baton Rouge, you informed him that you had told Under Secretary Brown that the 17th Street Canal levee had broken on Monday, August 29, and you received that information about 11 a.m. Was he surprised by this information? And did he indicate whether Under Secretary Brown had informed him of the levee breach on Monday?

Mr. BAHAMONDE. He told me when I gave him that information—he was very curious, and he said, "That doesn't fit with what I have already been told. I understood that the levee didn't break until Tuesday." And I said, "No, sir. I informed FEMA Headquarters at 11 a.m. on Monday that it broke, and I informed Under Secretary Brown at 7 o'clock that night that the levee had broke." And he was very curious with the information I was giving him because he said a lot of what I was telling him just didn't fit with the information that he had been provided.

Senator AKAKA. In your past experience doing advance work for Under Secretary Brown, were you the only person prepositioned at the location of a disaster or were you there accompanying an operations team or working with another agency?

Mr. BAHAMONDE. In other assignments or in this one in particular?

Senator AKAKA. Was there another group there in New Orleans?

Mr. BAHAMONDE. No. I was the only one in New Orleans.

Senator AKAKA. You stated in the Committee interview that a FEMA medical response team from Oklahoma arrived at the Superdome at 3 a.m. on Tuesday morning, and it was the first team to arrive on the scene. According to your statement, that team left the Superdome later that day when another medical response team from New Mexico arrived.

Do you know why the Oklahoma team left only hours after they had arrived?

Mr. BAHAMONDE. I do not know that.

Senator AKAKA. You mentioned that FEMA likes to surround the perimeter of a disaster and preposition people and supplies to go in after the storm passes. Could FEMA have tightened the perimeter? In other words, could they have been closer in the center of the disaster so they could have deployed more promptly?

Mr. BAHAMONDE. The closest medical team at the time was in Houston, which is about an 8-hour drive, I am estimating, a 6- to 8-hour drive. But I do know that there were members that were staged as close as Baton Rouge, which is only an hour and a half away. But I don't know what the decision was. I didn't make the decision or have any information as to why they were prepositioned all the way in Houston.

Senator AKAKA. Under Secretary Brown said that Mayor Nagin handed him a list of New Orleans' needs during their initial meeting and did not discuss the list at the meeting. However, you stated that Mayor Nagin verbally shared the list with the Under Secretary.

Can you share what was on Mayor Nagin's list, to the best of your recollection?

Mr. BAHAMONDE. I don't recall because when that meeting took place, I certainly wasn't part of the meeting. I didn't sit down and participate in it. There were a lot of things that I was trying to do, working with the Coast Guard and working with helicopters to try to get supplies in. So my modus operandi was not to be a part of that meeting but was to continue the work that I was there for.

Senator AKAKA. Under Secretary Brown testified that the person he sent down to New Orleans, which was you, had comparable qualifications to a Federal coordinating official. Do you feel you have the same operational background and qualifications as a Federal coordinating official?

Mr. BAHAMONDE. No. I certainly don't have the training. I haven't been through the courses. But I have field experience over 12 years, which I certainly know what I am doing out on a disaster, and I certainly know how they work.

Senator AKAKA. Under Secretary Brown stated in his testimony before the House that over the last few years FEMA had lost a substantial number of its career staff and that its budget had been raided to fund other programs in DHS. As a long-time employee, do you share that view? And if so, do you have any insights as to which programs have suffered?

Mr. BAHAMONDE. Well, I certainly know the staff that has left, the institutional knowledge that has left, has really hampered the agency because there were career employees there who have done this for 10, 15, 20 years that have seen a lot of disasters that are now gone, replaced by others who don't have emergency management experience and really don't have the experiences.

So within FEMA, it is well known and difficult to know that the institutional knowledge was either forced or chose to leave on their own out of the agency, and I think it has hampered us in very significant ways.

Senator AKAKA. You have provided a summary of events that largely focuses on interaction between FEMA and the local government. Can you share with us what interaction, if any, you observed between the State and local governments and what interaction you had with the State Government?

Mr. BAHAMONDE. I didn't really have any contact with the State Government when I arrived in New Orleans other than the National Guard, who was directed there by the State. I did attend some—there were some calls that were taking place in the New Orleans EOC that linked in both the State and FEMA and other agencies just to provide an update as to what was going on. So I do know that those existed, but I didn't have any direct contact with anybody in the State about any plans or operations that they were involved in.

Senator AKAKA. Well, Mr. Bahamonde, I really appreciate your responses and wish you well. Hafa Adai. Thank you.

Chairman COLLINS. Thank you, Senator. Senator Pryor.

OPENING STATEMENT OF SENATOR PRYOR

Senator PRYOR. Thank you, Madam Chairman.

I want to thank you for your candor and your transparency. I know that you have been asked lots of questions about this, and I just really appreciate the time and the detail you are giving us here.

Let me say that my questions are going to be a little bit of a scattershot because I am following up on a lot of my colleagues' questions here. But just by way of background, in your opening statement you said you have been with FEMA for 12 years?

Mr. BAHAMONDE. That is correct.

Senator PRYOR. But full-time since 2002.

Mr. BAHAMONDE. That is correct.

Senator PRYOR. How does that work? You were part-time before or called in as needed or what?

Mr. BAHAMONDE. FEMA has a disaster assistance employee cadre of thousands of people that are not full-time employees, that really only deploy during a time of a disaster or when other special assignments are required. I think that is the most unique and special aspect of FEMA, that we have people throughout the country that in their normal lives are wives, husbands, self-employed, work at a bank, or work at a hospital. But when a disaster is called, they drop everything and respond. And that is what I was for the first 8 years.

Senator PRYOR. So for those 8 years, what did you do in your normal life?

Mr. BAHAMONDE. I went back to school. I got a master's degree and pursued other opportunities.

Senator PRYOR. OK. So how many FEMA Directors did you serve under then?

Mr. BAHAMONDE. I have served under three.

Senator PRYOR. OK.

Mr. BAHAMONDE. Four now with Chief Paulison.

Senator PRYOR. OK. So those would include James Lee Witt——

Mr. BAHAMONDE. Correct.

Senator PRYOR. To start with, I guess you started in James Lee Witt's FEMA.

Mr. BAHAMONDE. Correct.

Senator PRYOR. I want to ask you about the changes you have seen in FEMA over time, over those 12 years. Could you describe in general terms some of those changes?

Mr. BAHAMONDE. I think in my early years, certainly during the 1990s, there were a lot of very proactive programs that were put in place. James Lee Witt was very instrumental in what I would say is thinking out of the box. Project Impact was a program that was started in the late 1990s, which was really designed to bring in corporations, get States and local governments involved. And it was just developing to be a very important and significant role in emergency management. When the administration changed, that was one of the very first programs that was cut and disbanded. And I know that a lot of employees in FEMA were dismayed to think that we were going in a different direction that didn't seem to be as forward thinking as we had just come from. And I know that agency-wide, speaking to employees, that was a significant moment that said that there were changes ahead.

Senator PRYOR. Tell me a little bit about Project Impact.

Mr. BAHAMONDE. I don't have a whole lot of experience with Project Impact, but I know that it was an effort to get corporations involved in working with local communities, working with State communities, to develop a well-rounded emergency management plan in and amongst those communities. So you get a major corporation to work within their community to develop emergency plans and a structure and use corporate money to help fund that. And it really brought in more than just the Federal Government. It brought in industry, private companies, and was really, what I envision, a ground-breaking moment to make everyone in this country involved in emergency management, not just government.

Senator PRYOR. And is it fair to say that some of your colleagues at FEMA were disappointed that Project Impact was cut?

Mr. BAHAMONDE. Oh, absolutely.

Senator PRYOR. And is it fair to say that some of your colleagues there felt it was a mistake to cut it?

Mr. BAHAMONDE. Absolutely.

Senator PRYOR. And do you share those views?

Mr. BAHAMONDE. Absolutely.

Senator PRYOR. And what about the morale at the agency? Quite frankly, here in the Senate, we hear sometimes about morale in various agencies, and we have heard some things about FEMA's morale. But I would like your thoughts on that if you could talk about it in general terms, please.

Mr. BAHAMONDE. I think if you go back to the end of the James Lee Witt administration, FEMA was described as the model agency in the Federal Government. There was tremendous pride in the agency. We wore our shirts, we wore our hats, we wore our jackets with pride because we knew that we were efficient, we did good work, and we made a difference in the lives of Americans.

Right now there is not very good morale in the agency, as things have been cut, programs that people had dedicated their lives to work on were all of a sudden gone. Staff has left. People that you could count on and turn to have gone. Training was virtually eliminated for some time period. And I think there was a general feeling that where the agency came from after Hurricane Andrew and the problems that the agency experienced then to rise to the top of the mountain and be proud of what you did, to where we stand today, I think it—you can't help but not have bad morale.

Senator PRYOR. Thank you for that candor, again. Now, as I understand it, you were working in External Affairs?

Mr. BAHAMONDE. Correct.

Senator PRYOR. Could you describe the activities in External Affairs?

Mr. BAHAMONDE. External Affairs has five components: Public affairs, congressional affairs—I am proud to say Senator Lieberman and Senator Collins are two of the Senators that I am responsible for in New England—international affairs, community relations, and State and local government.

Senator PRYOR. OK. Let me ask about the standard operating procedure at FEMA in relation to an expected hurricane. Is it standard operating procedure there to send one External Affairs person to a major city, in this case New Orleans? It seems to me that maybe there ought to be more of a team approach, but tell me how that should work and how that does work.

Mr. BAHAMONDE. There is an emergency response team that in this instance was deployed to Baton Rouge which did contain an External Affairs person with it as well as an operational person, a logistical person, and everything. But they were never sent to New Orleans. They were sent to coordinate out of Baton Rouge.

Senator PRYOR. Now, standard operating procedure, should they have been sent to New Orleans or was it the right call to send them to Baton Rouge?

Mr. BAHAMONDE. It is standing policy of FEMA not to put employees into harm's way, and that is why they were based in Baton Rouge and not in New Orleans.

Senator PRYOR. Unless you are a lone External Affairs guy, right?

Mr. BAHAMONDE. I guess so. [Laughter.]

Senator PRYOR. OK. But, is it standard operating procedure to send one person down into harm's way?

Mr. BAHAMONDE. No.

Senator PRYOR. OK. So why did you get sent down there?

Mr. BAHAMONDE. Because I think I have a long history with Under Secretary Brown. I believe I have his trust. I believe that he—I can confide in him and speak very frank to him, and he can communicate with me and know that he is going to get the straight story. So sending me to an area where I think he needs to know that he has a person there that he has the full trust and is confident in is the reason why I was being sent to these locations to provide that information.

Senator PRYOR. OK. It sounds as if you were being sent there maybe as eyes and ears, but you did not have a lot of authority within FEMA to direct resources?

Mr. BAHAMONDE. I was sent there as eyes and ears to develop an advance in the event that he—which was already planned—was going to visit New Orleans. Typically a storm hits somewhere, he comes in. They need to know where to go, what they are going to see, where a helicopter can land.

I know most of you are familiar with what kind of advance work takes place when VIPs make visits. That was my responsibility. But when I got down there, I realized that there was a desperate need for another operation, which I felt comfortable at some levels overtaking, speaking with the mayor, confiding in him as to what was going on, getting information, passing that up. I have done that in numerous disasters before and certainly felt comfortable in that situation, but that is not what my purpose was.

Senator PRYOR. It sounded to me like you provided a vital link and also some vital expertise there on the ground that was just desperately needed. Is that fair to say?

Mr. BAHAMONDE. Yes.

Senator PRYOR. And I could have a lot of follow-up questions on that, but let me change gears, if I may, just for one moment. You talked about the Superdome, and there has been a lot of focus on the Superdome, understandably. As I understand it, there always existed, even at the height of the flooding in New Orleans, there always existed roads in and out of the city, including in and out of the Superdome, that could have gotten supplies in and people out. Is that true?

Mr. BAHAMONDE. There were access points that you could get into. Certainly if you weren't familiar with the city and you weren't escorted by city police, you would have no way of knowing how to get that because it was a very circuitous route that took you over several bridges and around back roads and back in. And that was really the only route in and out of the city.

Senator PRYOR. But helicopters could have helped determine that as well, right? Helicopters could have provided the road map because they could have seen what was available.

Mr. BAHAMONDE. Absolutely.

Senator PRYOR. And how much time did you personally spend in and around the Superdome?

Mr. BAHAMONDE. Well, I spent most of Sunday afternoon and Sunday evening at the Superdome, and then I returned to the Superdome Monday evening for the helicopter tours and all that stuff. And then beginning Tuesday morning, I never left the Superdome for any significant amount of time until I left on Thursday.

Senator PRYOR. So you overnighted there?

Mr. BAHAMONDE. I slept with everyone else there.

Senator PRYOR. OK. For, what, two nights?

Mr. BAHAMONDE. Two nights.

Senator PRYOR. We have, again, been focusing on the Superdome. But also there was a whole scene going on at the Convention Center as well. Did you ever go to the Convention Center?

Mr. BAHAMONDE. No, I did not.

Senator PRYOR. Did you know what was going on in the Convention Center?

Mr. BAHAMONDE. Yes, to some extent on Wednesday morning, and the mayor had been saying as early as Sunday he was very concerned what was going to happen to 30,000 tourists that were in the city. He reiterated that again on Tuesday, and on Wednesday he reiterated again that hotels were starting to kick people out of their hotels. They no longer had food themselves. They didn't have power, they didn't have an infrastructure in place for the hotels to house these people. So they started moving them out into the streets. And so I sent an E-mail out on Wednesday morning explaining what that situation was and how it was developing.

Senator PRYOR. And were you getting reports from the Convention Center?

Mr. BAHAMONDE. No. I was getting reports from the mayor's staff.

Senator PRYOR. OK. Now, again, to change gears here just a moment, one of the fundamental questions I think you brought up in your opening statement was whether FEMA should fundamentally be a response organization or be a recovery organization.

Mr. BAHAMONDE. Response is—we are not a response agency in the respect that we don't have the infrastructure that the military has, we don't have the helicopters, we don't have some of the trucks and stuff like that. We respond with personnel, we respond with communications equipment, and we respond with logistics. But we are not a first responder in a local community during a crisis.

Senator PRYOR. OK. Should you be?

Mr. BAHAMONDE. That is not how we are designed, but certainly with more assets and a lot more tools at our disposal, I think that would be left up to Congress to decide if they want that to happen or not.

Senator PRYOR. I am going to change gears on you again. One thing you mentioned in responding to questions a few moments ago is unreliable information coming from FEMA. Were you receiving any reliable information from FEMA?

Mr. BAHAMONDE. Could you clarify that? What do you mean "unreliable information"?

Mr. BAHAMONDE. Apparently at one point you were told that trucks were coming with X number of MREs, and it was significantly less than that. Was that information coming from FEMA, or who was that coming from?

Mr. BAHAMONDE. The information on the food, the MREs, that was coming from the National Guard. The information that I was getting from FEMA was specifically about the medical team, where they were, how far they were out, if they were going to arrive. That is the first information that I received there. So it was accurate information because it was happening. They just never got there.

Senator PRYOR. In Arkansas, we took in a lot of people from Louisiana and Mississippi and a few from Alabama, and we found that the information we were receiving from FEMA was very unreliable, about how many were coming, when they would arrive, what their condition was. And then there were reports here yesterday and the day before about an ice truck coming out of the Northeast and heading south and then ending back up in Massachusetts. It just appears to us that communication within FEMA at a very critical time was lacking. Was that your experience?

Mr. BAHAMONDE. I don't think anybody is going to dispute that once commodities and things are left for others to deposit, we a lot of times lose visual sight as to where they are and where they are going and if they even get there. We can put the ice and the water and the food on a truck. Sometimes we don't always know when it is going to arrive and where it is going to arrive.

Senator PRYOR. The last thing I had is you mentioned that you did some media interviews.

Mr. BAHAMONDE. Correct.

Senator PRYOR. What was your message? Was this before or after the storm?

Mr. BAHAMONDE. It was before the storm.

Senator PRYOR. What was your message?

Mr. BAHAMONDE. Well, it was just basically to talk about where our commodities were, where we had stationed our assets. I talked about the perimeter, that we had put a perimeter really around the Gulf Coast, both in Florida, Alabama, and Mississippi, Louisiana and Texas, so that depending upon when the storm hit, we could move in any direction.

As it turned out, the problem with that plan was the only direction we could move in was from the west, so any assets over on the east side I assume went to Mississippi, which I was not familiar with. But it was just generally that conversation, conversation about whether FEMA was more worried about terrorism and less worried about natural disasters. I remember specifically commenting on that. But it was just a general, 3- or 4-minute interview.

Senator PRYOR. Thank you.

Thank you, Madam Chairman.

Chairman COLLINS. Thank you. Senator Lautenberg.

OPENING STATEMENT OF SENATOR LAUTENBERG

Senator LAUTENBERG. Thanks very much, Madam Chairman.

Mr. Bahamonde, when I think of the role you played, it takes me back a little bit in history when someone hollered, "The British are coming, the British are coming," and you were hollering, "The waters are coming, the waters are coming." And I give you the mantle of a modern-day Paul Revere. I think what you did was heroic, and we are grateful to you for your persistence in trying to get your message across.

Mr. BAHAMONDE. Thank you.

Senator LAUTENBERG. Some of the things that are so perplexing are when did the information come to the White House. I mean, did it come to the White House later than these photos? They are hard to see from here, but this is Monday afternoon, August 29, there are two people swimming. And I remember this one, taken from television, of people walking with bundles on their heads, some of them with children on their shoulders, because the water was chest-high and they were afraid of what might happen. And we know that on occasion terrible things did happen and children slipped out of the hands of their parents.

So here these pictures are being presented to the public at large—to the world at large, as a matter of fact, and it sounds like someone at the White House may have been scratching their head and saying, well, how important is this?

I ask if you have any idea when the White House might have heard from Mr. Brown or perhaps from Secretary Chertoff. Do you know when the President might have gotten notice that the situation was so desperate?

Mr. BAHAMONDE. I don't. All I know is when I spoke to Under Secretary Brown at approximately 7 o'clock on Monday evening, the only thing he said to me was, "Thank you. I am now going to call the White House."

Senator LAUTENBERG. That was Monday night?

Mr. BAHAMONDE. Correct.

Senator LAUTENBERG. About what time?

Mr. BAHAMONDE. Around 7 o'clock.

Senator LAUTENBERG. I can't imagine what would have happened if we had a few thousand of our troops encircled by enemy forces someplace in the world, whether or not the President might have continued his vacation, as he did—and I am not asking for your opinion; I am making a statement—whether or not he would continue with his vacation plans, not feeling the urgency to get to the situation room at the White House, to say the least, and give some direction to this. I mean, how could anybody look at these pictures and not be so motivated to run and—to use an analogy—grab the fire extinguisher and go do something about it?

So it is a tragedy that occurred, and we know that "Brownie" didn't deserve a pat on the back; he deserved a kick. You described what you were told by Mr. Brown, "to get your ass back into New Orleans." So the ineptitude was obvious. Why he still represents the fall guy in this thing—he was bad, got fired, he still continued to get paid for a while, and I think got another assignment. But to suggest that he alone should have taken the blame for all the failures that we saw—and, believe me, there was heroism throughout from ordinary citizens, from your colleagues at FEMA, from the Coast Guard.

When I saw the Coast Guard—and I am a fan of the Coast Guard and always have been. I remember the picture so vividly with the guy being pulled up on a rope line or whatever line they used to get him off the roof. And why aren't the fingers being pointed with the same aggression, with the same anger, with the same rage, the fact that on Tuesday—Tuesday—when this thing was practically at its height, the President was in California with a guitar in his hand celebrating an occasion? No one will ever understand why that White House wasn't surrounded by black cloth and saying, hey, let's get on with this, let's do whatever America can do. It has not happened.

What was the level of experience of your colleagues, senior management positions? In a generalization, were they people who were well trained for this kind of thing? Because you were there as a professional, and I want to understand something. To an earlier question, you responded that you were part of the Department of Public Information. But yet you had been through these other very serious hurricanes. What was your job at those moments?

Mr. BAHAMONDE. Well, in some of the other hurricanes, I did typical advance work. I certainly worked with Under Secretary Brown at the time, did advance work with Secretary Ridge, and I have worked on several occasions with the White House's advance team.

But my primary responsibility is public affairs, external affairs.

Senator LAUTENBERG. Were you doing that in New Orleans on August 29?

Mr. BAHAMONDE. Doing what?

Senator LAUTENBERG. Public affairs?

Mr. BAHAMONDE. I was because I did some media interviews.

Senator LAUTENBERG. But the calls, some of them of a relatively frantic nature in terms of your appeal for food, for oxygen——

Mr. BAHAMONDE. Just because I am a public affairs person——

Senator LAUTENBERG. Wasn't that kind of an operational assignment?

Mr. BAHAMONDE. Well, I cannot just sit and stand by and see what is happening and not let somebody know. If you stay in your box, people would die. And so I have been on a lot of disasters. I see what happens. And I took it upon myself to let everybody that I could possibly know what the situation was so that they would be able to make informed decisions on what to do next.

Senator LAUTENBERG. Didn't you adopt the position of communications director effectively on August 29?

Mr. BAHAMONDE. I wouldn't say communications director. I would say I was all you had——

Senator LAUTENBERG. Was there anybody else——

Mr. BAHAMONDE [continuing]. And you should listen.

Senator LAUTENBERG. Anybody else? You weren't called out of a bull pen to do this. I mean, you were there. Was there anyone else from FEMA in New Orleans on that Sunday?

Mr. BAHAMONDE. No.

Senator LAUTENBERG. Monday?

Mr. BAHAMONDE. No.

Senator LAUTENBERG. You took on enormous responsibility, and it goes beyond public affairs. You were the chief life guard, and it is a big job. We are grateful to you.

Thursday morning, September 1, President Bush stated on "Good Morning America," "I don't think anyone anticipated the breach of the levees." Is there anything that you saw in documents, reviews of the strength of the levees while at FEMA in advance of this where people were aware of the fact that levees might be in a less perfect condition than they should be to withstand a heavy rush of water like that, which finally evolved?

Mr. BAHAMONDE. Well, I don't have any personal knowledge of the strength and the knowledge of the levees. I do know that FEMA has worked with the State and the city for that inevitability. We had done extensive training for that fact.

Senator LAUTENBERG. So information was exchanged about——

Mr. BAHAMONDE. The worst-case scenario, as long as I have been in the agency, was a levee break to flood the City of New Orleans. So the possibility——

Senator LAUTENBERG. How long ago would you say that data came to you, how long ago?

Mr. BAHAMONDE. Almost as long as I have been in the agency.

Senator LAUTENBERG. You are there 12 years?

Mr. BAHAMONDE. Yes.

Senator LAUTENBERG. It is too bad that President Bush didn't think that anyone anticipated the breach of the levees because it sounds like a certain detachment that I think is appropriate for a commander in chief. I remember when the President flew over on Wednesday and thought that it was a devastating sight. And I compared it to the unacceptable behavior of wearing white shoes after Labor Day. It was a casual reference to what a devastating picture.

Anyway, I think that you have done a service for your country that equals many of the moments of greatness, and we thank you for it.

Thank you, Madam Chairman.

Chairman COLLINS. Thank you.

Senator Carper, you have been extraordinarily patient all morning to get your questions in. It is my understanding, however, that our witness needs a short break, and I wonder if you would be amenable to a 5-minute break before we proceed with your questions?

Senator CARPER. Given all that he has been through, he can take 10 minutes. [Laughter.]

Mr. BAHAMONDE. Thank you, Senator.

Chairman COLLINS. We will stand in recess for 5 minutes.

Mr. BAHAMONDE. Thank you.

[Recess.]

Chairman COLLINS. The Committee will come to order.

We will now resume questioning with the ever patient Senator from Delaware, Senator Carper.

OPENING STATEMENT OF SENATOR CARPER

Senator CARPER. Well, Mr. Bahamonde, I would like to say we have saved the best for last, but my colleagues would say that is not the case.

I want to echo the sentiments of others who have expressed our thanks for your service, your remarkable service, and not just in this instance but obviously with others.

Coming last, this has some advantages, and one of the advantages is having listened to all the other questions and to have the opportunity to listen to your responses to them. And I think I would like to use part of this time, at least initially, to take advantage of hindsight. We always say hindsight is 20/20. Mine is—I am much better at hindsight than I am in looking ahead. I expect that is true for most of us.

Before I ask a question along those lines, I want to first ask, if you will, what are some of the principal or primary lessons that you would have us as Members of this Committee, as Members of the Senate, take away from this hearing and from your testimony?

Mr. BAHAMONDE. I would certainly ask you to look at where emergency management is at this time, what you expect the true role of FEMA to be, because I think at some point certainly the employees have gone through the last couple of years uncertain themselves as to what our true role is. There has been a lot of focus on terrorism, and that is expected. But we know within the agency we do 50, 60 natural disasters a year. We certainly need to be upgraded and brought into the 21st Century and the technology that we have, how we do communications, how we deploy teams. But I think what you need to do is to define what you want the role of FEMA to be, where you want it to be, how you want it to operate, what command you want it to have. And once that is defined, then I think FEMA has a tremendous amount of assets staff-wise, career people, who desperately want your guidance and desperately want to get back to the job that they know how to do and have proven in the past to be very successful and just want to go back home, so to speak.

Senator CARPER. When you think of the assets that FEMA brings to a disaster like this, what are the major assets?

Mr. BAHAMONDE. Communications is something that we bring a lot to. But I think one thing that was obvious to me at the Superdome is all of our communications is on a truck or in some sort of camper. I would have certainly loved to have a little briefcase that the National Guard had that you can pop up a satellite dish and throw open a phone and automatically have communication. We didn't have any of that in the Superdome. We were communications dead. We begged and borrowed and pleaded with the National Guard to let us use their communications, because we didn't have any. And I kept standing there looking at their little satellite dish that they just deployed and thought, Wow, that would be cool to have.

So I think it is those kind of assets that, as we look into the future on how we are going to have to respond, both in a natural disaster and a terrorist event, you have to do it immediately, and you have to have the tools in which you allow the employees to do the work that they are trained to do.

Senator CARPER. Going back to what I said earlier about hindsight being 20/20, knowing what we know now, what could we have done, should we have done differently in preparing for this disaster, which was well telegraphed, and responding to it?

Mr. BAHAMONDE. Yes, and I think what we saw——

Senator CARPER. That is a pretty big question. I just want you to think about it and just try to tell us what could we have done differently in preparing for this disaster and in responding to it? And not just with a focus on FEMA, but you had the opportunity to observe other agencies, certainly the Coast Guard, the Corps of Engineers, to observe the National Guard, to observe the response of State and local people. But what could we collectively have done differently and should have done differently in preparing for this disaster and responding to it?

Mr. BAHAMONDE. I will go to money because I think when this disaster is all said and done, there is talk that it is going to cost $200 billion to recover from this disaster, and I would just like to think if we just had a couple of billion dollars to put some of the plans, some of the equipment, some of the preparation, strengthening the levees certainly around the city that you have known for as long as it has existed it is underwater, put plans into evacuation. We saw what evacuation did in Hurricane Rita in getting people out of harm's way.

If we had had that in New Orleans, we wouldn't be here today.

Senator CARPER. Anything you want to add to that?

Mr. BAHAMONDE. I can talk a lot, but——

Senator CARPER. I have 9 minutes. [Laughter.]

Mr. BAHAMONDE. There are just a lot of things there—but I will leave it at that.

Senator CARPER. All right. You mentioned communications, and we have had a fair amount of discussion within this Committee and in the Senate itself on communications breakdowns. I am an old governor, and we had invested a lot of our own resources, Delaware State resources, in 800 megahertz communications systems that are interoperable and allow all of our first responders—police, fire, paramedics, National Guard, and others, DEMA personnel—to be able to communicate, pretty much from one end of the State to the other.

In terms of communications, this is the first that I personally heard someone talk about the kind of technology that the National Guard was able to deploy and the advantage that it gave them.

Do you have any further thoughts with respect to communications that go beyond just FEMA and their communications capabilities, or lack thereof?

Mr. BAHAMONDE. Well, certainly, when the City of New Orleans went down, they lost all their communications. So I think you have to look at it from a wide-ranging aspect. Are our cities in a position to communicate when a disaster hits? We have the fortune, in the nature of a hurricane, to know what is going to happen. But in the event of a terrorist attack, do we have the infrastructure, do we have the capability—and I can't believe in the 21st Century that the technology doesn't exist that we can't acquire that.

As you know, once communications go down, there is a cascading effect where everything else goes down. And I think we had that problem in New Orleans, and I think we paid the price for it.

Senator CARPER. If the levees had held, if the flood walls had not been breached, would we be here today?

Mr. BAHAMONDE. Oh, absolutely. I think there were still thousands of people in the city. There was a Catagory 5 hurricane coming. There was a lot of damage. It wasn't just caused by water. But it certainly played a significant role.

But, yes, anytime you have a Catagory 5 hurricane—and I certainly hope in the future that nobody ever stays around to witness one, and I think we learned that lesson, and hopefully with Hurricane Wilma out there, we will continue to learn that lesson.

Senator CARPER. You have had an opportunity to be involved up close and personal with any number of natural disasters, I sense—hurricanes——

Mr. BAHAMONDE. Yes.

Senator CARPER. Could you just compare and contrast for us the response of the State and local folks, including the Guard, local police, local officials, State officials, their response, the effectiveness of their response to that you have observed in other instances?

Mr. BAHAMONDE. Certainly there's different levels of response, but I think the critical aspect is you have to have a plan, you have to know your plan, and you have to have the ability to carry out your plan. It is impossible to develop a plan and try to implement the plan when it is most critical to carry out that plan. And some cities and some States do a phenomenal job because they have invested the time, the energy, and the money to do it. Other States and local governments haven't, for whatever reason. But I know the Department of Homeland Security has spent billions of dollars over the last several years trying to arm first responders, trying to arm local communities, emergency management groups with money to buy technology resources and do it. And it really comes to the responsibility of local officials and State officials to take that money and use it as it is designed, to create the plans and put in place the implementation, whether it be resources or staff, to carry those out. And until we actually come to that realization, we may be faced with more situations like this.

Senator CARPER. Talk with us here, if you will, about the specific role that the National Guard played. My understanding is that FEMA usually does not preposition assets in harm's way, instead showing up shortly after the storm or whatever the natural disaster is, once the worst has passed. Talk with us about the role of the National Guard, and particularly with respect to this storm. When they showed up, when they were called up, were the numbers of men and women adequate? Were they prepared for what faced them?

Mr. BAHAMONDE. I thought the National Guard at the Superdome did a fabulous job. They made sure that everybody who went into the Superdome was inspected so that it was a safe environment. They made sure there weren't weapons or drugs or paraphernalia. They hand-inspected everything that people brought in there to create a safe environment. They were there before, they were there during. They did yeoman's work. They had communications. And it still amazes me today that you put 25,000 people into the Superdome under the conditions that they were living under, and that there was relatively few incidents of the crowd getting unruly, and I credit that to the National Guard. I think they had a firm grip on what was going on there, and I commend them for the work they did at the Superdome.

Senator CARPER. Thank you.

Share your thoughts with us, looking ahead, about whether or not we ought to try to rebuild New Orleans. To what extent should we go forward and in some areas maybe hold back? What is an appropriate Federal role in your judgment?

Mr. BAHAMONDE. That is not for me to say. That is not my position.

Senator CARPER. OK. I want to come back to some more specific questions, if I could. We talked earlier when my colleagues asked you about food and water, particularly that was provided to folks in the Superdome. Let me just ask, the food and water that folks were able to find for Superdome evacuees, any idea where it came from?

Mr. BAHAMONDE. FEMA had prepositioned some of it, and then when they had an opportunity, drove it into the theater of New Orleans so that it was easier accessed, if need be.

Senator CARPER. OK. At any time during your stay in the Superdome—what were you, 2 or 3 days? Did you say 2 days?

Mr. BAHAMONDE. At the Superdome?

Senator CARPER. Yes.

Mr. BAHAMONDE. Where I physically stayed?

Senator CARPER. Yes.

Mr. BAHAMONDE. Yes, from Tuesday morning until Thursday.

Senator CARPER. Were you aware of any FEMA employees who were working as you were to get food or water or more supplies to the Convention Center—to either the Superdome or the Convention Center?

Mr. BAHAMONDE. Absolutely. I know of one incident where a FEMA employee went above and beyond the call of duty to make sure that food got to that Superdome, and he did it in the dark of night, almost by himself. And so he is—he did a great job.

Senator CARPER. OK. Senator Akaka had some questions about a medical team. I think there were two medical teams—one from Oklahoma and maybe one from New Mexico. As I recall, one was deployed, showed up, and then another one came along. Was it the Oklahoma medical team that was pulled out?

Mr. BAHAMONDE. That was my understanding, but I am not——

Senator CARPER. Any idea where they were sent to?

Mr. BAHAMONDE. I really don't know specifically where that team was sent to. I know other medical teams were stationed primarily out at New Orleans airport to set up a base there.

Senator CARPER. OK. I want to go back to the first question that I asked and just revisit that since my time is about to expire. If you were in our shoes, not your shoes, but if you were in our shoes, just delineate again for me and for my colleagues what are the next several things that you would do to reduce the likelihood that this kind of—I am tempted to call it a tragedy, to reduce the likelihood that this is going to happen again.

Mr. BAHAMONDE. And I would go back to you. If the Federal Emergency Management Agency is the Federal Government's overseeing body to natural disasters and terrorism, you need to define what their role is. They were placed under the Department of Homeland Security, and they are playing a role there. But it played a much different role in many respects outside of the agency. And speaking from just an employee perspective, I just want—it would be helpful to know what you want us to do, where you want us to go, and what our future is so that—because right now there are a lot of employees who are debating whether to stay or not because the future is cloudy and they are not sure if they want to commit to that anymore.

Senator CARPER. Thank you. Thank you again for your testimony here and thank you for your remarkable service to our country and to the people of New Orleans.

I would just say to my colleagues, one of the thoughts that comes to mind, during our short recess a few minutes ago, I discussed this with Senator Lieberman. We have responsibilities in this, too. You know, we decide whether or not FEMA is going to be part of Homeland Security or not. We decide with the rest of our colleagues the level of funding that is appropriate and some idea what their mission is to be.

We also have some say in who we confirm to provide leadership for FEMA and for other organizations. And I am struck as we consider the role that FEMA played in Louisiana and the reviews, not very good reviews they have gotten for the most part, compared to the role that the Coast Guard played and have gotten just remarkably excellent reviews.

But as one Member of this Committee, I am going to be taking a lot more seriously my responsibilities not just for oversight, but my responsibilities to look carefully at the credentials and qualifications of those that are nominated to serve in these senior leadership positions to make sure that they bring to the job the skills that are needed. Thank you very much.

Chairman COLLINS. Thank you.

Mr. Bahamonde, the end is almost in sight. I am sure you are happy to hear that, but you really have been extraordinarily valuable in enhancing our understanding this morning.

Usually your E-mails and phone calls from New Orleans went to officials at FEMA other than Michael Brown, the Director. But you did actually call Mr. Brown and direct some E-mails personally to him. I assume that was because you wanted to make sure that he personally was aware of how dire the situation was in New Orleans. Is that correct?

Mr. BAHAMONDE. That is correct.

Chairman COLLINS. Now, Senator Warner brought up the issue that you informed Mr. Brown but that you did not specifically ask him to take certain steps or to send in more help or to provide additional supplies. And just for the record, I want to clarify that issue.

Did you not ask or make specific requests of him because you felt that the information you were providing was sufficient to trigger those kinds of actions?

Mr. BAHAMONDE. Absolutely. I was not in the position to make the demands or the commands. I was in a position to provide information. But I did say, "Sir, we need food at the Superdome. We need water at the Superdome. We need medical teams at the Superdome." I did say that.

Chairman COLLINS. And to me that should have been more than sufficient to trigger an urgent and more effective response, but I did want to just clarify that point for the record.

The last issue—because I see a roll call vote has just started—that I want to cover with you has to do with Secretary Chertoff's proposals to change what I would call the eyes and ears function that you performed in New Orleans with respect to future disasters. Yesterday he testified before the House that he intended to put emergency response teams on the ground to provide real-time information to headquarters and to improve situational awareness in the field. That is the kind of role that you were playing on the ground in New Orleans, correct?

Mr. BAHAMONDE. It turned out to be, yes.

Chairman COLLINS. What do you think of his proposal to preposition emergency response teams to take over that eyes and ears role that you played?

Mr. BAHAMONDE. I think it is critical. I think you need that. I think you have to be there. But if you don't take and use the information you are provided with, it means nothing.

Chairman COLLINS. That is an excellent point, and I think that is one of the lessons from this hearing.

Finally, have you gotten any feedback from FEMA or DHS as to the role you performed? Have you participated in any after-action review of what went wrong and what went right with FEMA's response?

Mr. BAHAMONDE. No, no one has talked to me about it.

Chairman COLLINS. Thank you. Senator Lieberman.

Senator LIEBERMAN. Thanks, Madam Chairman.

Mr. Bahamonde, I want to go back to something we discussed earlier, which is when FEMA officials, and Michael Brown in particular, learned of your reports that the 17th Street Canal levee had broken Monday morning, you said you discussed the 11 a.m. report you overheard in the city Emergency Operations Center with Mr. Brown's special assistant Mike Heath. Is that right?

Mr. BAHAMONDE. That is correct.

Senator LIEBERMAN. But you also said you don't know whether Mike Heath passed that information on to Michael Brown, right?

Mr. BAHAMONDE. Correct.

Senator LIEBERMAN. Just to make the record complete, I want to share with you—and I know the clerk has just given it to you—an E-mail chain that we received just last night from FEMA. And I apologize to you and the other Members of the Committee that this did not make it into the exhibit book. But I think it is important to raise and include in the record now.[1]

In this chain, as you will see in the document I have given you, which is a copy of a chain, it is an 11:51 a.m. E-mail on Monday, August 29, from Mike Heath to someone named Michael Lowder. Who is Michael Lowder? Do you know the name?

Mr. BAHAMONDE. I believe he is the Deputy Director of Operations? Is that correct?

He is one of the senior leaders in the response operation.

Senator LIEBERMAN. OK, within FEMA.

Mr. BAHAMONDE. Correct.

Senator LIEBERMAN. This E-mail says, "From Marty."—you—"He has been trying to reach Lokey."—who I know is another FEMA official who was in Baton Rouge at the time. "New Orleans FD"—I presume "Fire Department"—"is reporting a 20 foot wide breech [sic] on the [L]ake [P]onchartrain [sic] side levy [sic]. The area is [L]akeshore Blvd and 17th Street."

Then at 11:57 a.m., again, still Monday morning, Mr. Lowder forwarded this on to Michael Brown with a note that says, "Not sure if you have this."

At 12:08 on that Monday morning, Mr. Brown replied, "I'm being told here water over not a breach"—in other words, that the water has gone over the levee but the levee hasn't been breached or broken.

To which Mr. Lowder quickly replied, "Ok. You probably have better info there. Just wanted to pass you what we hear."

Now, I am not sure if you want to comment on that, but I did want to let you know that your message actually did make it to Mr. Brown within a relatively short time after you overheard the comment.

Mr. BAHAMONDE. Good to see.

[1] Exhibit submitted by Senator Lieberman appears in the Appendix on page 68.

Senator LIEBERMAN. OK. Just to add a few things that make this so troubling, so puzzling. We know that you previously testified to the evening direct conversation with Michael Brown, telling him what you had seen by that time after you had gone up in the Coast Guard helicopter, levee is breached—levees are breached, 80 percent of New Orleans underwater. Senator Levin earlier quoted from the MSNBC 9 p.m. interview with Michael Brown that night, Monday, August 29. "Just beginning to receive reports from my folks in the field of literally tens of square miles of homes inundated with water up to the roofs."

Incidentally, I presume anybody following this hearing appreciates this fact, but I want to just go back and restate it. You are not just somebody who coincidentally was dropped into New Orleans. Mr. Brown clearly trusted you. He dispatched you based on experiences that you had together. And he goes on national television—and unless he heard it from somebody else, but I don't believe there was anybody else he could have heard it from—he immediately repeats what you told him about the flooded conditions. And then Mr. Brown said, "I have already told the President tonight that we can anticipate a housing need of at least in the tens of thousands."

Secretary Chertoff yesterday appears before the House committee, and he says, "When I learned on Tuesday morning that the challenge had been compounded because there had been an irreparable levee breach or a series of levee breaches," it is just amazing to me that after all that you had done during all of Monday in a critical situation where the water is literally rising, somehow the word does not reach Mr. Chertoff until Tuesday morning. And then he says, "At that point I recognized"—he said this yesterday in the House committee—"that we needed to be—perhaps be somewhat more—have somewhat more intense focus in terms of what was being done in the field that might have been done in the case of a normal hurricane."

He also said yesterday—and, of course, we are going to want to ask him this—that he was trying unsuccessfully to reach Michael Brown all day Tuesday, didn't reach him until Tuesday at 8 p.m., which is a puzzle because you obviously reached him by phone—Mr. Brown—and you were E-mailing directly during the day.

The other troubling part—maybe it was a misstatement, I don't know—September 2, the following Friday, it would have been, President Bush is in Biloxi, Mississippi, after his first visit to the Gulf Coast, and he says, "The levees broke on Tuesday in New Orleans"—not on Monday, as we know they did. It has worked its way up.

This has unsettling echoes in a very different context of the September 11 tragedy in the sense that information was in different places, in this case particularly prior to the attack, and it wasn't reaching the key decisionmakers in a coordinated way for them to take action and prevent it.

Maybe the other metaphor, which is a painful one, is that as we now know, the first responders in the World Trade Center couldn't communicate with each other because of the inability of their systems to do so, and it looks in a very different context like a repeat of that.

I want to ask you a question similar to the one that Senator Collins did, but in a slightly different way. You have already testified that you felt, based on what you saw at the Superdome, beginning Sunday, in which you were telling your superiors that they should have sent several medical teams to deal with it and emergency response teams. What in your mind, as you first heard on Monday morning, August 29, that the levees had been broken, and as the day went on saw that the city was 80 percent flooded, what in conveying that up did you hope would happen in response?

Mr. BAHAMONDE. That they would bring everything that the Federal Government had at its disposal to help those people.

Senator LIEBERMAN. Amen. Thank you very much for your testimony today. You have really built a factual basis here from firsthand eyewitness experience about what was happening, what you tried to do, and we have to continue to follow this trail where the facts take us to figure out why it was not done and what is being done to make sure that nothing like this happens again. So you really deserve a medal for what you did then and your willingness to come forward and talk today, and I thank you very much for both.

Mr. BAHAMONDE. Thank you for the opportunity.

Chairman COLLINS. Mr. Bahamonde, usually I do a closing statement, but there is an E-mail that you sent, which is Exhibit 22,[1] which to me sums up why we are having this investigation, the stakes, and why this matters so much. And so instead of my making a closing statement today, I would like to ask you to read the E-mail that is Exhibit 22.

Mr. BAHAMONDE. I wrote this E-mail to Cindy Taylor on Saturday, September 3, at 1 o'clock in the morning, and I said, "The State told us"—it was in reference to planes who were no longer able to fly evacuees out because there were such delays that the pilots needed rest, and she ended her E-mail by saying, "I'm bitter, I'm frustrated, I'm angry." And I wrote her back, and I said, "The State told us we would run out of places. The Army told us they had run out of places. The leadership from top down in our agency is unprepared and out of touch. When told that the Superdome had been evacuated and would be locked, Scott Wells said, 'We shouldn't lock it because people might still need it in an emergency.' Myself and the general who had been there immediately spoke up and told Scott that it was impossible to send anyone back in there. Is that not out of touch? But while I am horrified at some of the cluelessness and self-concern that persists, I try to focus on those that have put their lives on hold to help people that they have never met and never will. And while I sometimes think that I can't work in this arena, I can't get out of my head the visions of children and babies I saw sitting there helpless, looking at me and hoping I could make a difference. And so I will and you must, too. It is not what we do that is as important as who we are, and that's what those little kids' faces were counting on."

Chairman COLLINS. I thank you so much for your testimony today and for your courage in coming forward to tell your story.

[1] Exhibit No. 22 appears in the Appendix on page 00.

This hearing record will remain open for 15 days, and this hearing is now adjourned.

Mr. BAHAMONDE. Thank you.

Chairman COLLINS. Thank you.

[Whereupon, at 12:30 p.m., the Committee was adjourned.]

APPENDIX

Testimony of Marty J. Bahamonde
Office of Public Affairs, Federal Emergency Management Agency
before the
Senate Committee on Homeland Security and Governmental Affairs
Thursday, October 20, 2005, 10:00 a.m.

Good Morning. I appreciate the opportunity to appear before you today.

I am Marty Bahamonde. I work for the Federal Emergency Management Agency (FEMA) as a Public Affairs Officer for FEMA' s Boston office and worked in FEMA's Headquarters in Washington, D.C. I worked in New Orleans prior to and immediately following Hurricane Katrina and have spent that past 6 weeks working at the Joint Field Office in Baton Rouge. I was the only FEMA employee deployed to New Orleans prior to the storm. I am here today to tell you what I experienced during the five days before and after Hurricane Katrina in New Orleans and specifically at the Superdome.

Let me briefly explain why I was there.

Since 2003, I have often been tasked by the Undersecretary of FEMA Mike Brown and his staff to do advance work for the Undersecretary in preparation for or responses to large disasters. My assignments included Hurricane Isabel in 2003, Hurricane Charley in 2004, Hurricane Dennis this past June and others. My responsibilities varied, but always included providing accurate and important information to FEMA's front office and Under Secretary Mike Brown. On Friday, August 26th, I was tasked by FEMA's front office to work advance wherever Hurricane Katrina was going to hit, which by Saturday appeared to be New Orleans.

I arrived in New Orleans Saturday night, August 27. Through the generosity of the New Orleans Emergency Operation Center (EOC), I was able to work in their office, and they provided me with shelter during the hurricane. The city EOC is located in City Hall, almost directly across from the Superdome.

On Sunday, August 28, I met the city staff at the EOC, got to know people and their roles at the EOC, and developed my own plans for my advance work, which included coordinating with the Coast Guard to arrange a flyover after the Hurricane passed. On Sunday, Katrina intensified to a Category 5 storm. I sensed a great deal of worry among everyone. I was worried too. My contact at FEMA Headquarters told me to leave New Orleans because it would be too dangerous. But like the thousands of other people left in New Orleans, the traffic jams leaving the city that morning prevented me from leaving.

The Superdome had been opened as a special needs shelter, but on Sunday, as thousands of residents were unable to evacuate, the Superdome became a shelter of last resort for anyone left in the city. By noon, thousands began arriving and by midday, lines wrapped around the building. It was also at that time that I realized that the size of the crowd was a big concern at the EOC. Terry Ebbert, the city's Homeland Security Director, made an announcement in the EOC that struck me. He asked the maintenance staff to gather up all of the toilet paper in city hall and any other commodities they could find and immediately take them over to the Superdome. I specifically note this because it told me that supplies at the dome might be a serious issue.

I was between the Superdome and the EOC throughout the day on Sunday. I took pictures of the crowds and sent them back to FEMA Headquarters. On Sunday evening, I was at the Superdome to do media interviews and afterwards I met with the National Guard inside the Superdome to discuss a range of things including the expected arrival of a FEMA Disaster Medical Assistance Team (DMAT) from Houston. The National Guard also told me that they expected 360,000 "meals ready to eat" "MRE's" and 15 trucks of water to arrive that night. As the storm intensified outside, a series of critical events began to unfold. Instead of 360,000 MRE's, only 40,000 arrived. Instead of 15 trucks of water, only 5 arrived, and the medical team did not arrive either.

Later that night after most of the 12,000 evacuees entered the Superdome, I returned to the EOC around midnight to ride out the storm. By early Monday morning, with the storm upon us, reports from throughout the city were moderately optimistic; some low level flooding, no levee breaks and limited wind damage. But by 8 a.m., the nearest point of eye passage, the situation worsened, I could clearly see and reported back to FEMA Headquarters that the Hyatt hotel and other tall buildings in downtown had suffered incredible window damage, and I could see the roof peeling off the Superdome. I received several calls from FEMA Headquarters seeking confirmation of the situation on the ground.

At approximately 11 a.m., the worst possible news came into the EOC. I stood there and listened to the first report of the levee break at the 17th Street Canal. I do not know who made the report but they were very specific about the location of the break and the size. And then they added it was "very bad". I continued to provide regular updates to FEMA Headquarters throughout the day as the situation unfolded.

At approximately 5 p.m., I rushed over to the Superdome because I had been notified that a Coast Guard helicopter was able to take me for a short flyover so that I could assess the situation in the city and plan for Under Secretary Brown's visit the next day. My initial flyover lasted about 10 minutes and even in that short time I was able to see that approximately 80 percent of the city was under water, and I confirmed the 17th Street Canal levee break. I was struck by how accurate the 11 a.m. call was about the levee.

About 15 minutes later, I went back up on a second Coast Guard helicopter for approximately 45 minutes, and during this flight, I was able to get a real understanding of the impact of Katrina on New Orleans and the surrounding area.

Upon landing, I immediately made three telephone calls. The first was to Under Secretary Mike Brown at approximately 7 p.m. The second was to FEMA's front office, and the third was to FEMA Public Affairs. That third call was to set up a conference call with FEMA Operations in Headquarters, the Emergency Response Team-National (ERT-N) team in Baton Rouge, the Regional Response Coordination Center (RRCC) in Denton, Texas, and with FEMA's front office, so I could make as many people aware of the situation that faced FEMA and the City of New Orleans.

In each report and on the conference call, I explained what I saw and then provided my analysis of what I believed to be the most critical issues we were facing:

- Ground transportation into the city was virtually non-existent because of the massive flooding. Any ground transportation must come from west of the city because the I-10 bridge to Slidell on the east side of the city was completely destroyed and there was no access from the north because of flooded roads. I also stated that the situation would only worsen in the next day or so because of the massive amounts of water being deposited into the city via the levee break. I described the levee break as being really bad.
- Search and rescue missions were critical as thousands of people stood on roofs or balconies in flooded neighborhoods.
- Supplying commodities would be a challenge as more and more people were headed to the Superdome to escape the flood waters and food and water supplies were already very short at the Superdome. I told them that the Superdome population was at almost 20,000 people.
- Medical care at the Superdome was critical because the staff there had run out of oxygen for special needs patients and more and more people needed medical attention.
- Housing an entire city worth of people would be a major issue as approximately 80 percent of the city was underwater to varying degrees and many areas were completely destroyed.
- Environmental issues would be major as I reported that an oil tanker had run aground and was leaking fuel.

I believed at the time and still do today, that I was confirming the worst case scenario that everyone had always talked about regarding New Orleans.

I then had a meeting with New Orleans Mayor Nagin and his staff of approximately 25 people, and I told them of the situation so that they would know what they faced in the hours and days ahead. It was a very emotional meeting for everyone. Following that meeting, Terry Ebbert pulled me aside and said, "You have done this before, what do we need to do now?" I told him that he needed to make a list of the city's priorities, actions and commodities, so that they could move forward with an organized plan. I also told him that they needed to let the State know so that FEMA could be tasked to help. He said, "Consider it done."

On Tuesday, August 30, I woke early to discover that water was rising around the Superdome and that by 6 a.m., 2-3 feet of water was in the streets around City Hall and the Superdome, and it was continuing to rise. I called Bill Lokey, the Federal Coordinating Officer in Baton Rouge immediately and told him of the rising waters and once again tried to express the seriousness of the situation. He told me he would have an operational team deployed to the Superdome later that day to relieve me.

About an hour later I was informed that Under Secretary Brown, Governor Blanco, Senators Landrieu and Vitter and others were planning on flying to the Superdome later that morning. At approximately 8 a.m., I went to find Mayor Nagin, who was at the Hyatt hotel, to inform him of the visit and asked him to meet with the group. So I put on tennis shoes and shorts and walked through what was now waist deep water over to the Hyatt, and told the Mayor of the visit. I then walked through the water again to get to the Superdome.

At the Superdome, I spoke with the National Guard to get the latest conditions and it was obvious that the Superdome conditions were in rapid decline and that there was a critical need for food and water. I communicated this to Under Secretary Brown when he arrived later that morning. I told him that the Superdome conditions were deplorable, and that we desperately needed food and water.

During the subsequent meeting with the Mayor and the incoming group, Mayor Nagin pulled out his list of priorities and proceeded to tell everyone what he needed for his city and the residents.

By early Tuesday morning a FEMA Medical Team arrived at the Superdome and by early afternoon a four-member Emergency Response Team- Advance (ERT-A) arrived. For the next three days I worked and lived at the Superdome with the ERT-A team and with the FEMA Medical Team. Each day it was a battle to find enough food and water and get it to the Superdome. It was a struggle, meal-to-meal, because as one was served, it was clear to everyone that there was not enough food or water for the next meal. But because of some truly heroic efforts from FEMA staff, the Coast Guard, and the National Guard, enough food was always found and brought to the Superdome so that at least two meals were served each day.

While we battled food and water issues, rescue missions continued, more and more people arrived at the Superdome, and the medical conditions of many at the Superdome were in rapid decline, many people were near death. Working in absolutely deplorable conditions, the FEMA New Mexico DMAT Team saved lives. They worked with helicopter Medivacs to evacuate the most critically ill.

I would like to say that what has been lost in all of the discussions and criticisms of what did or did not happen in New Orleans, is that real heroes existed there and the FEMA New Mexico medical team did truly amazing things to save lives and make a difference. They worked 24 hours a day to treat patients and it can be said that without their work, more people certainly would have died. I am honored that I can call them co-workers because they are the best of what FEMA has to offer.

It is well known what happened over the next several days, most of it real, some of it hyped and exaggerated by the media, but all of it tragic. I am most haunted by what the Superdome became. It was a shelter of last resort that cascaded into a cesspool of human waste and filth. Imagine no toilet facilities for 25,000 people for five days. People were forced to live outside in 95-degree heat because of the horrid smell and conditions inside. Hallways and corridors were used as toilets, trash was everywhere, and amongst it all children, thousands of them. It was sad, it was inhumane, and it was so wrong.

By Thursday, plans to evacuate the Superdome were underway, but there were critical missions to still carryout at the Dome. But early Thursday, the National Guard approached the FEMA staff and told us that there was intelligence that a riot was being planned for noon and that they did not want to be responsible for our safety so they recommended that we leave. Phil Parr, who was the senior FEMA official on the ground, made the decision to evacuate all FEMA assets from the Dome including the medical team. I strongly voiced my concerns about abandoning the mission and the critical need to continue with medical care and the coordination of food and water into the Dome. I pointed out that overnight, approximately 150 heavily armed forces arrived at the Superdome by helicopter raising the security level. I called Mike Brown to tell him that we were leaving. I contacted FEMA's front office to let them know that we were leaving. Within an hour, all FEMA personnel were ordered onto trucks and driven out of the Dome. Our leaving meant that FEMA lost visibility of the situation and operational control at the Superdome. I do not believe that it was the right decision for us to leave.

I have worked for FEMA for 12 years and have been a full-time employee since 2002. I have spent most of that time in the field, not behind a desk. I have responded to numerous hurricanes, earthquakes, floods, tornadoes and fires. At the personal request of Under Secretary Brown, I deployed to Bam, Iran in 2003 in support of the medical team that worked miracles in a city that was totally destroyed and resulted in more than 30,000 deaths. I have seen the worst mother nature can hand out, and I saw it in New Orleans and at the Superdome.

My purpose before you today is to help you understand what happened in New Orleans and when it happened, as I know it from my own perspective. I hope that what you learn from me and the many others that will come later, is a better understanding of emergency management and response. From this I hope that we are able to effect change, so that no other child, no other senior citizen, no other special needs patient, no other parent, and no other community in this country will have to experience the horrors and tragedy that happened in New Orleans and the entire Gulf Coast.

I will now answer any questions that you may have.

Questions For the Record

Senate Homeland Security & Governmental Affairs Committee
"Hurricane Katrina in New Orleans: A Flooded City, A Chaotic Response"
October 20, 2005
Marty Bahamonde, Regional Director, External Affairs, Region One
Federal Emergency Management Agency

Questions from Senator Ted Stevens

1) In 1999 the Appropriations Committee directed FEMA, in the Department of Veterans Affairs, Housing Development and Independent Agencies Appropriations Bill of 2000, to develop an evacuation plan for a category 3 or greater storm, a levee break, flood or other natural disaster in the New Orleans area. Can you tell me if this plan was ever developed?

Response: Attached is a history of previous hurricane study projects.

The Congressional direction in H.R. 106-286 provided no funding above the annual program funding for the National Hurricane Program (NHP) for the development of an evacuation plan, which is a State and local responsibility.

FEMA, through its NHP, has undertaken hurricane evacuation studies in 22 States, Regional areas and territories. Through these studies, NHP develops the products and tools for State and locals to conduct evacuation planning and operationally to conduct the actual evacuations. NHP develops the technical information that is used by State and local governments to develop an evacuation plan. This includes information such as the evacuation clearance time, transportation analysis, hazard analysis, and behavioral study. The evacuations and the planning involved are a State and local responsibility using the tools and products developed by the FEMA. The budget for the NHP is $2.91 million per year, and approx. $2.9 million goes directly to the States through the grants.

For the New Orleans area including the surrounding Parishes a hurricane evacuation study was completed in 1994 and used as the basis for the hurricane evacuation plans. A hurricane evacuation restudy was scheduled for completion in FY 2006 (Southeast LA Hurricane Evacuation Restudy (HES)). Some preliminary information from this restudy was provided prior to Hurricane Katrina. This information included a preliminary surge analysis and draft clearance times for evacuation. The Southeast LA Hurricane Evacuation Restudy information is currently being reviewed by the State, FEMA, and the USACE (the HES Study Manager) in consideration of the effects of Hurricane Katrina.

Of interest, the FY 2006 Homeland Security Appropriations Conference Report (109-24) directed the Secretary of DHS (through the Office of Domestic Preparedness) to report on "...catastrophic planning, including mass evacuation planning in all 50 states and 75 of the largest urban areas by February 10, 2006. The report should also include certifications from each state and urban area as to the exact status of plans for evacuations of entire metropolitan areas in the state and the entire state, the date such plans were last updated, the date exercises were last conducted using the plans, and plans for sustenance of evacuees".

Questions For the Record

Senate Homeland Security & Governmental Affairs Committee
"Hurricane Katrina in New Orleans: A Flooded City, A Chaotic Response"
October 20, 2005
Marty Bahamonde, Regional Director, External Affairs, Region One
Federal Emergency Management Agency

2) As the only FEMA official in New Orleans in advance of Hurricane Katrina, can you tell me what role FEMA played in implementing and coordinating the evacuation of the city?

Response: FEMA has no statutory authority to order the evacuation of local jurisdictions and although it was not directly involved in evacuation efforts before Hurricane Katrina made landfall in New Orleans, FEMA did maintain contact with Evacuation Liaison Teams prior to landfall. Traditionally, and as assumed in the National Response Plan (NRP), State and local emergency response plans should make provisions for and identify the resources needed for evacuations and protection of the public. Protecting their citizens through evacuations and other measures is the responsibility of the elected/appointed officials and this responsibility is codified in State/local laws and statutes. State/local emergency response plans should clearly lay out timing parameters and roles and responsibilities associated with carrying out evacuations, account for special needs populations, and account for people who do not have their own means of transportation. States and local jurisdictions often enter into interstate and intrastate agreements/compacts to share resources in evacuations.

FEMA can assist in an evacuation by providing overtime reimbursement to local and State jurisdictions under the Public Assistance Program (Category B - Emergency Protective Measures) when the President declares an Emergency or Major Disaster Declaration under the Stafford Act.

The NRP designates the Department of Transportation (DOT) as the lead support agency to DHS/FEMA for Emergency Support Function 1 (ESF-1)/Transportation, and calls for DOT to do the following during an Incident of National Significance:

1) Process and coordinate requests for Federal and civil transportation support;
2) Report damage to transportation infrastructure as a result of the incident;
3) Coordinate alternate transportation services;
4) Coordinate the restoration and recovery of transportation infrastructure;
5) Perform activities conducted under the direct authority of DOT elements such as air, maritime, surface, rail and pipelines; and
6) Coordinate and support prevention/preparedness/mitigation among transportation infrastructure stakeholders at the State and local levels.

ESF-1 emergency support staff began responding to Hurricane Katrina well before the storm made landfall. On August 24, 2005, the ESF-1 staff in Washington, D.C. and in Region 4, which includes Florida, Alabama, Mississippi and several other southeastern states, was put on alert. On August 27, 2005, ESF-1 staff in Region VI, which includes Louisiana and Texas, was alerted for activation in the Regional Response Coordination Center on August 28, 2005.

Questions For the Record

Senate Homeland Security & Governmental Affairs Committee
"Hurricane Katrina in New Orleans: A Flooded City, A Chaotic Response"
October 20, 2005
Marty Bahamonde, Regional Director, External Affairs, Region One
Federal Emergency Management Agency

Between August 31 and September 5, 2005, FEMA/DHS coordinated with DOT to evacuate nearly 70,000 people by bus to sites across the U.S. In another example of evacuation support, on August 31, 2005, FEMA's Logistics Transportation Unit tasked Emergency Support Function 1 (DOT) to acquire 455 buses and 300 ambulances through ESF-1. More specifically, on August 31, 2005, a tasker issued as part of an existing mission assignment called for 455 buses and 300 ambulances to be sourced by DOT beginning August 31, 2005, to evacuate people from the New Orleans Superdome. DOT did provide 455 buses on August 31 and an additional 150 on September 1 for a total of 605 buses.

The request to DOT for ambulances was subsequently cancelled and directed to GSA to source and provide the ambulances. On September 1, 2005, the original tasker was amended to provide even more buses bringing the total to 1,105 buses and another amendment to the tasker on September 3, 2005, added another 250 buses bringing the total to 1355 buses provided by DOT. However, this number was subsequently decreased back to 1,105. Because indications were that the Superdome evacuees had departed by September 03, 2005, subsequent buses were also used to pick up evacuees elsewhere in the City. An AMTRAK train was also made available through the efforts of USDOT's Federal Rail Administration to support the evacuation from New Orleans. Overall, there was close and effective coordination between FEMA and DOT relative to sourcing and providing busses and no significant issues were encountered.

Questions For the Record
Senate Homeland Security & Governmental Affairs Committee
"Hurricane Katrina in New Orleans: A Flooded City, A Chaotic Response"
October 20, 2005
Marty Bahamonde, Regional Director, External Affairs, Region One
Federal Emergency Management Agency

Attachment

History of Hurricane Related Study Products
State of Louisiana

1. Greater New Orleans Expressway Commission Master Plan Phase I (1998)
   Congressional earmark $ 500,000 through FEMA and administered by LOEP (Various contractors)
   (Study regarding feasibility of Lake Ponchartrain Causeway expansion/options)

   Phase II: 2000
   Congressional earmark $500,000

   Final Product: **"Comprehensive Master Plan and Execution for Transportation System and Hurricane Evacuation Design Year 2016"**

2. Hurricane Lili (2002) Post Storm Assessment (Released 2003)
   FEMA, US ACE, LOEP
   Contractor: PBS & J

3. Southeast Louisiana Catastrophic Hurricane Plan (2005)
   FEMA, LOSHEP (formerly LOEP)
   Contractor: IEM
   Catastrophic planning funds from FEMA Headquarters

History of Hurricane Related Study Products
State of Louisiana

1. Greater New Orleans Expressway Commission Master Plan Phase I (1998)
   Congressional earmark $ 500,000 through FEMA and administered by LOEP (Various contractors)
   (Study regarding feasibility of Lake Ponchartrain Causeway expansion/options)

   Phase II: 2000
   Congressional earmark $500,000

   Final Product: **"Comprehensive Master Plan and Execution for Transportation System and Hurricane Evacuation Design Year 2016"**

2. Hurricane Lili (2002) Post Storm Assessment (Released 2003)
   FEMA, US ACE, LOEP
   Contractor: PBS & J

3. Southeast Louisiana Catastrophic Hurricane Plan (2005)
   FEMA, LOSHEP (formerly LOEP)
   Contractor: IEM
   Catastrophic planning funds from FEMA Headquarters

Committee on Homeland Security and Governmental Affairs
EXHIBIT #1

Committee on Homeland Security and
Governmental Affairs
EXHIBIT #2

Committee on Homeland Security and Governmental Affairs
EXHIBIT #3

Committee on Homeland Security and
Governmental Affairs
EXHIBIT #4

Committee on Homeland Security and Governmental Affairs
EXHIBIT #5

Committee on Homeland Security and Governmental Affairs

EXHIBIT #6

Bahamonde, Marty

From: Bahamonde, Marty
Sent: Sunday, August 28, 2005 4:40 PM
To: 'Michael.Heath1@dhs.gov'
Subject: Re:

There are currently 2000 inside the dome with about 300 special needs. National guard is setting up cots. Oklahoma-1 DMAT is expected at around midnight to supply medical needs. Medical staff at the dome say they expect to run out of oxygen in about two hours and are looking for alternatives.
There are still lines outside the dome waiting to get in

CG has confirmed they will pick me up at the superdome at first light Tuesday morning weather permitting

Contact: 202-841-7750
617-272-5149

Sent from my BlackBerry Wireless Handheld

Committee on Homeland Security and Governmental Affairs
EXHIBIT #14

Taylor, Cindy

From: Taylor, Cindy
Sent: Monday, August 29, 2005 1:38 PM
To: Rule, Natalie; Widomski, Michael; Andrews, Nicol D - Public Affairs
Cc: Passey, David; Hudak, Mary
Subject: Updates from Marty and Beeman

From Marty (New Orleans):

- 17th Avenue Canal levee (running along border of Orleans and Jefferson Parishesbroke through with water flow "bad" into New Orleans side
- City will have feeding/sheltering issues with population at superdome (est. 12K)
- North side of city under est. 11' water in heavy residential area
- Estimate of 30,000 tourists in city holed up in hotel rooms
- Charity Hospital - all windows out; basement flooded, no power; have not been able to reach other 27 hospitals for capabilities
- Early estimate that power will be over a month to restore - main hubs out of commission
- Hoping to get out in next hour, with helicopter tour to follow

From Beeman (Gulfport):

- Last broadcasting radio station was knocked off the air about an hour ago
- Significant flooding occuring not far from where he's located with reports of people climbing into attics to escape waist deep and higher waters
- Been sitting at desk for last four hours watching as the winds have torn apart the back of the building across, with pieces of the roof slamming into the posts that hold up the porch on his buildings.
- Trees dropping all around and six foot fence from a small park to left have been sailing by like frizzbies
- Power and water are both off

Committee on Homeland Security and Governmental Affairs
EXHIBIT #22

Taylor, Cindy

From: Bahamonde, Marty
Sent: Saturday, September 03, 2005 1:06 AM
To: Taylor, Cindy
Subject: Re: Airlift

The state told us we would run out of places, the army told us they had run out of places. The leadership from top down in our agency is unprepared and out of touch. When told that the superdome had been evacuated and would be locked, Scott Wells said we shouldn't lock it becuase people might still need it in an emergency. Myself and the general who had been there immediately spoke up and told scott that it was impossible to send anyone back in there....is that not out of touch. But while I am horrified at some of the cluelessness and self concern that persists, I try to focus on those that have put their lives on hold to help people that they have never met and never will. And while I sometimes think that I can't work in this arena, I can't get out of my head the visions of children and babies I saw sitting there, helpless, looking at me and hoping I could make a difference and so I will and you must to. It is not what we do that is as important as who we are and that's what those little kids faces were counting on.

Sent from my BlackBerry Wireless Handheld

Committee on Homeland Security and Governmental Affairs
EXHIBIT - Sen. Lieberman

From: Lowder, Michael [Michael.Lowder@dhs.gov]
Sent: Monday, August 29, 2005 12:09 PM
To: 'Michael.D.Brown@dhs.gov'
Subject: Re: Information

Ok. You probably have better info there.

Just wanted to pass you what we hear.

-----Original Message-----
From: Brown, Michael D <Michael.D.Brown@dhs.gov>
To: 'Michael.Lowder@dhs.gov' <Michael.Lowder@dhs.gov>; 'Michael.D.Brown@dhs.gov' <Michael.D.Brown@dhs.gov>
Sent: Mon Aug 29 12:08:37 2005
Subject: Re: Information

I'm being told here water over not a breach

-----Original Message-----
From: Lowder, Michael <Michael.Lowder@dhs.gov>
To: 'Michael.D.Brown@dhs.gov' <Michael.D.Brown@dhs.gov>
Sent: Mon Aug 29 11:57:35 2005
Subject: Fw: Information

Not sure if you have this..

-----Original Message-----
From: Heath, Michael <Michael.Heath@fema.gov>
To: Lowder, Michael <Michael.Lowder@fema.gov>
Sent: Mon Aug 29 11:51:18 2005
Subject: Re: Information

FYI...

From Marty. He has been trying to reach Lokey.

New Orleans FD is reporting a 20 foot wide breech on the lake ponchatrain side levy. The area is lakeshore Blvd and 17th street.

Sent from my BlackBerry Wireless Handheld

-----Original Message-----
From: Lowder, Michael <Michael.Lowder@fema.gov>
To: Brown, Michael D <Michael.D.Brown@fema.gov>; Lokey, William <William.Lokey@fema.gov>; Jones, Gary <Gary.Jones@fema.gov>; Robinson, Tony <Tony.Robinson@fema.gov>; Heath, Michael <Michael.Heath@fema.gov>; Rhode, Patrick <Patrick.Rhode@fema.gov>; EST-ESF03-A <EST-ESF03-A@fema.gov>; EST-ESF03-B <EST-ESF03-B@fema.gov>; EST-ESF03C <EST-ESF03C@fema.gov>
Sent: Mon Aug 29 09:52:34 2005
Subject: FW: Information

ML

———

From: Green, Matthew
Sent: Monday, August 29, 2005 9:50 AM
To: Buikema, Edward
Cc: Lowder, Michael
Subject: RE: Information

From WWL TV

.. A LEVEE BREACH OCCURRED ALONG THE INDUSTRIAL CANAL AT TENNESSE STREET. 3 TO 8 FEET OF WATER IS EXPECTED DUE TO THE BREACH...LOCATIONS IN THE WARNING INCLUDE BUT ARE NOT LIMITED TO ARABI AND 9TH WARD OF NEW ORLEANS.

Matthew Green
FEMA Hurricane Liaison Team Coordinator
National Hurricane Center
11691 SW 17th Street
Miami, Florida, 33165-2149 USA
305-225-4217

———

From: Buikema, Edward
Sent: Monday, August 29, 2005 9:40 AM
To: Green, Matthew
Cc: Lowder, Michael
Subject: Information

Matthew,

Please copy me on the spot report emails you send to Mike Lowder. That information is very helpful.

Thanks. Ed

APEX
TP
9780160768309
76901953 – 17